中等职业教育烹饪专业教材

烹饪厨房英语

(第 2 版)

主　编　张　毅　贾颖丽
副主编　车　霞　朱　玉　关　华
参　编　南　丽　窦　莉　周　岩　高　冲

重庆大学出版社

内容提要

本书是一本极具特色的烹饪厨房英语教材。本书图文并茂，形象、生动地介绍了烹饪厨房英语的词汇。以对话的形式，将厨师间的会话、餐厅点菜等内容包括其中；以短文阅读的形式，介绍了国际酒店标准、菜肴的制作过程、饮食习惯、餐饮文化等内容；以课后练习的形式，帮助学生灵活运用所学的句型和词汇，提高其口语表达能力。本书共16个单元，其中，第11单元至第14单元专为烹饪专业不同方向的学生设置，满足了烹饪专业不同方向学生的学习需要。本书针对性强，内容丰富，难度适中，情景交融，生动形象，是一本实用性很强的教材。

本书可供已完成基础英语学习的烹饪专业学生学习，也可作为烹饪从业人员学习和培训的教材，还可供广大烹饪爱好者自学。

图书在版编目（CIP）数据

烹饪厨房英语 / 张毅，贾颖丽主编. -- 2版. -- 重庆：重庆大学出版社，2021.10（2023.1重印）
中等职业教育烹饪专业教材
ISBN 978-7-5624-8903-0

Ⅰ.①烹… Ⅱ.①张…②贾… Ⅲ.①烹饪—英语—中等专业学校—教材 Ⅳ.①TS972.1

中国版本图书馆CIP数据核字（2020）第164855号

中等职业教育烹饪专业教材
烹饪厨房英语（第2版）
主　编　张　毅　贾颖丽
副主编　车　霞　朱　玉　关　华
参　编　南　丽　窦　莉　周　岩　高　冲
责任编辑：沈　静　　　版式设计：博卷文化
责任校对：刘志刚　　　责任印制：张　策

*

重庆大学出版社出版发行
出版人：饶帮华
社址：重庆市沙坪坝区大学城西路21号
邮编：401331
电话：（023）88617190　88617185（中小学）
传真：（023）88617186　88617166
网址：http://www.cqup.com.cn
邮箱：fxk@cqup.com.cn（营销中心）
全国新华书店经销
重庆长虹印务有限公司印刷

*

开本：787 mm×1092 mm　1/16　印张：9　字数：280千
2015年6月第1版　2021年10月第2版　2023年1月第9次印刷
印数：20 501—23 500
ISBN 978-7-5624-8903-0　定价：45.00元

本书如有印刷、装订等质量问题，本社负责调换
版权所有，请勿擅自翻印和用本书
制作各类出版物及配套用书，违者必究

第2版前言

　　本书自2015年出版至今已有5年。5年来，本书受到了全国中等职业学校烹饪专业师生的喜爱。由于各方面反映很好，因此对本书进行了修订。

　　本书以实用性为编写原则，按照原料的种类和烹饪专业的不同方向将本书分为16个单元。本书内容丰富、难易适中、图文并茂，是一本实用性很强的烹饪厨房英语教材。本书符合职业教育办学规律，紧密结合行业的实际，能够引导学生全面进入烹饪、酒店等领域，将语言教学和职业教育融为一体，得到了中等职业学校烹饪专业师生的认可，被国内开设了烹饪专业的中等职业学校广泛使用。

　　为了使本书更加完善，更加实用，我们在原书的基础上进行了修订。修订内容如下：

　　1.对第1版中不恰当的地方进行了调整和删减。

　　2.降低了部分单元中阅读的难度，使学生更容易理解。

　　3.对不常见的词汇加以更新和替换，使其更加实用。

　　本书由张毅、贾颖丽担任主编，负责全书的统稿和部分章节的编写；车霞、朱玉、关华担任副主编，负责部分章节的编写和初稿的修改；南丽、窦莉、周岩、高冲担任参编。其中，贾颖丽编写Unit1，Unit2，Unit3，Unit15，Unit16，车霞编写Unit5，Unit6，Unit12，南丽编写Unit8，Unit9，Unit10，窦莉编写Unit13，Unit14，周岩编写目录，高冲编写Unit4，Unit7，Unit11。

　　在本次修订的过程中，我们收集了部分师生在教材中提出的宝贵意见，得到了一些热心朋友的大力帮助，也得到了重庆大学出版社和编者所在学校的大力支持，在此表示衷心的感谢！

　　尽管我们在修订过程中付出了不懈努力，但由于水平有限，书中可能仍存在一些不尽如人意的地方，敬请广大读者批评指正。

编　者
2021年7月

第1版前言

职业教育从本质上讲就是就业教育。中等职业学校的专业英语教学,应紧密结合行业实际,强调英语实际应用能力,提高学生的岗位适应能力和职业能力,使学生能够将学到的知识应用到行业中去,成为社会需要的有用人才。

随着我国旅游业的蓬勃发展和国际交流的进一步加强,饭店业得到了前所未有的发展和壮大,对饭店从业人员的英语交流能力的要求逐步提高。目前,中等职业学校的英语教学包括基础英语和专业英语两个部分,但专业英语的教材大都和餐饮服务有关,真正的烹饪厨房英语教材较少。为了提高中等职业学校烹饪专业学生的岗位应用能力,我们组织具有丰富教学经验的教师编写本书。本书不仅涉及原料的名称等相关内容,而且增加了烹饪方法、厨房设备和厨房工具、明档等方面的章节,以满足学生们的工作需要。本书符合职业教育办学规律,紧密结合行业的实际,引导学生全面进入饭店服务业领域,将语言教学和职业教育融为一体,体现了学以致用的原则。

本着实用性的编写原则,本书将烹饪厨房英语按照原料的种类和专业的不同分为16个单元,图文并茂、内容广泛、可读性强,许多材料和图片直接取材于酒店的后厨。

本书由张毅、贾颖丽担任主编,车霞、朱玉、关华担任副主编,南丽、窦莉、周岩和高冲担任参编。

本书在编写过程中,参阅了大量国内已经出版的有关资料,限于篇幅,我们没有一一注明出处,主要参考书目附于书末。我们希望以此表达对这些编著者的诚挚谢意。我们还要感谢大连教育学院职业学校教育中心的于红老师,她为本书的编写提出了许多宝贵的意见和建议。

由于编者水平有限,书中难免出现不足之处,敬请广大读者批评指正。

编　者
2015年2月

Unit 1 I Am a Cook 我是一名厨师 ·················· 1

Unit 2 Fruit 水果 ························· 8

Unit 3 Vegetable 蔬菜 ····················· 15

Unit 4 Meat & Poultry 肉和禽类 ················ 22

Unit 5 Seafood 海鲜 ······················ 30

Unit 6 Dairy & Egg Products 乳制品和蛋类制品 ········ 40

Unit 7 Drinks 饮品 ······················· 47

Unit 8 Dried Foods 干货 ···················· 54

Unit 9 Cooking Methods 烹饪方法 ··············· 61

Unit 10 Kitchen Equipment and Tools 厨房设备和厨房工具 ··· 69

Unit 11 Chinese Pastry 中式面点 ················ 77

Unit 12 Western Pastry 西式面点 ················ 84

烹饪厨房英语

Unit 13 Chinese Cuisine 中餐 ·· 92

Unit 14 Western Cuisine 西餐 ··· 99

Unit 15 Show Kitchen 明档 ·· 109

Unit 16 Hotel 酒店 ··· 118

Words ·· 127

参考文献 ··· 136

Unit 1

I Am a Cook
我是一名厨师

International Hotel Standard
国际酒店标准
Kitchen cooks should have good working knowledge of the English language.
厨师应该具有良好的英语工作知识。

Part A: Self-introduction

Good morning, everyone!

I'm Jack, 18 years old. I come from Dalian.

Now, I'm learning cooking in a cooking vocational school. My major is Chinese Cooking. I love my major very much!

My dream is to be a good cook in a five-star hotel. In order to make my dream come true, I will do my best to study hard.

Thank you!

Words & Expressions

cook [kuk] *n.* 厨师　　　　　　major ['meidʒə] *n.* 专业

hotel [həu'tel] *n.* 酒店　　　　vocational [vəu'keiʃənl] *adj.* 职业的

dream [dri:m] *n.* 梦，梦想

Notes:

1. My major is Chinese Cooking. 我的专业是中餐。 Chinese Cooking 中餐
   Western Cooking 西餐　　Chinese Pastry 中式面点　Western Pastry 西式面点
2. cooking vocational school 烹饪职业学校
3. five-star hotel 五星级酒店

Part B: Personal Hygiene

I'm not sick.

My fingernails are short and clean.

I wear hat to keep hair away from food.

My uniform/apron is clean.

I will wash my hands before preparing food.

I will wash my hands after:

   using the toilet

   sneezing

   blowing my nose

   handling raw food

   taking out the trash

Unit 1 I Am a Cook 我是一名厨师

Words & Expressions

sick [sik] *adj.* 恶心的，生病的
wear [weə] *vt.* 穿着，戴着
apron ['eiprən] *n.* 围裙
sneeze [sni:z] *vi.* 打喷嚏
raw [rɔ:] *adj.* 生的，未加工的
handle ['hændl] *vt.* 用手操作，处理或负责

fingernail ['fiŋgəneil] *n.* 指甲
uniform ['ju:nifɔ:m] *n.* 制服
toilet ['tɔilət] *n.* 厕所，洗手间
blow [bləu] *vt.* 吹，吹气
trash [træʃ] *n.* 垃圾，废物

Notes:

1. I wear hat to keep hair away from food. 我戴帽子防止头发掉入食品。
   这里的keep... away from... 为（使）……不接近，避开
   例如：Keep the children away from the machine. 别让孩子们接近机器。
2. blow my nose 擤鼻涕
3. take out the trash 扔垃圾

Part C: Dialogues

Dialogue 1

Tom: Hi, my name is Tom.
Jack: Hi, Tom. Nice to meet you. My name is Jack.
Tom: Nice to meet you too, Jack. Where are you from?
Jack: I'm from Harbin, Heilongjiang Province. What about you?
Tom: I'm from Dalian. What's your major?
Jack: My major is Chinese Cooking. How about you?
Tom: My major is Western Cooking. I love cooking very much.
Jack: I also love cooking very much. I believe I will be a famous cook in the future.
Tom: That's for sure.

Words & Expressions

province ['prɔvins] *n.* 省份
believe [bi'li:v] *vt.* 相信

famous ['feiməs] *adj.* 著名的，出名的

Notes:

1. I'm from Harbin, Heilongjiang Province. 我来自黑龙江省的哈尔滨。
2. That's for sure. 肯定会。

3

 烹饪厨房英语

Dialogue 2

Jack: Good morning.

Chef: Good morning.

Jack: What a nice day today!

Chef: Yes, let's start work now.

Jack: Just wait a minute. I'll go to wash my hands first.

Chef: OK. Hurry up! Do you know how to wash your hands properly?

Jack: Ahh ...

Chef: Let me tell you. First, wet your hands with clean running water and use soap. Second, soap your hands. Third, lather hands together to make a lather. Then, scrub all surfaces. Next, rinse hands well under running water. Finally, dry your hands using a paper towel or air dryer.

Jack: I see.

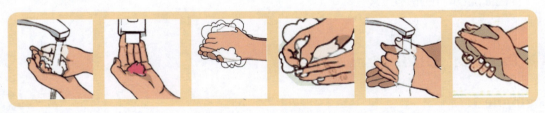

1. wet hands   2. soap   3. lather   4. scrub   5. rinse   6. dry

Words & Expressions

wet [wet] adj. 湿的；vt. 把……弄湿
lather ['lɑːðə(r)] n. 肥皂泡；vt. 用皂沫覆盖
rinse [rins] vt. 漂洗，冲洗
towel ['taʊəl] n. 毛巾
air dryer 空气干燥器
chef [ʃef] n. 厨师长

soap [səʊp] n. 肥皂；vt. 抹肥皂
scrub [skrʌb] vt. & vi. 用力擦洗，刷洗
dry [draɪ] vt. 从……去掉水分，使……变干
running water 流动的水

Notes:

1. What a nice day today! 今天天气真好！此句为由what引导的感叹句。
2. Just wait a minute. 稍等一下。
3. Hurry up! 快点！
4. First... second... third... then ... next ... finally...（表示顺序）第一，第二，第三，然后，接下来，最后……

Unit 1 I Am a Cook 我是一名厨师

International Hotel Standard:
国际酒店标准：
Kitchen cooks should know how to wash their hands properly before handing food to avoid contamination.
厨师应该知道如何在工作中正确地洗手，以免污染食品。

Exercises

Ⅰ. Match the correct words.

1. five-star hotel   A. 擤鼻涕
2. apron   B. 扔垃圾
3. blow your nose   C. 生食
4. sneeze   D. 洗手
5. use the toilet   E. 厨师
6. raw food   F. 围裙
7. province   G. 五星级酒店
8. wash hands   H. 打喷嚏
9. take out the trash   I. 上厕所
10. cook   J. 省份

Ⅱ. Complete the following dialogue and read it aloud.

> start   up   morning   nice   wash   minute

Jack: Good ____1____, teacher.
Teacher: Good morning, Jack.
Jack: What a ____2____ day today!
Teacher: Yes, let's ____3____ our study now.
Jack: Just wait a ____4____.
     I'll go to ____5____ my hands first.
Teacher: OK. Hurry ____6____!

Ⅲ. Translate the following sentences into English.

1. 你的专业是什么？
2. 我的专业是中餐/西餐/中式面点/西式面点。
3. 我的梦想是将来在一家五星级酒店当一名好厨师。
4. 我的指甲又短又干净。

 烹饪厨房英语

5. 擤完鼻涕后要洗手。
6. 我喜欢烹饪。
7. 快点！

Ⅳ. How to wash your hands:

First, _____ your hands with clean running water and use soap.
Second, _____ your hands.
Third, _____ hands together to make a lather.
Then, _____ all surfaces.
Next, _____ hands well under running water.
Finally, _____ your hands using a paper towel or air dryer.

Ⅴ. Write a self-introduction.

Part D: Reading

International Hotel Standard of Kitchen Cook

1. Kitchen cooks should have a pleasant and professional manner.
   厨师应该有愉悦、职业的态度。
2. Kitchen cooks should always be well groomed.
   厨师应该有良好的仪容仪表。
3. Kitchen cooks should be dressed in clean, pressed and complete uniforms.
   厨师应该穿着干净、熨烫平整的工作制服。
4. Kitchen cooks should appear organized and work as a team.
   厨师应该表现得有组织、有团队合作精神。
5. Kitchen cooks should greet guests and colleagues in a polite and friendly manner.
   厨师应该礼貌、友好地问候宾客和同事。
6. Plate covers should be clean and polished.
   餐盘盖子应该洁净、光亮。
7. Kitchen cooks should prevent cross contamination.
   厨师应该预防交叉感染。
8. Kitchen cooks should know the rule of FIFO（first in, first out）.
   厨师应该了解先进先出的原则。
9. Kitchen cooks should know the temperature danger zone is 4～60℃.
   厨师应该了解4～60℃为危险温度区。

Unit 1 I Am a Cook 我是一名厨师

10. Kitchen cooks must watch out for warnings.
    厨师必须注意警示标志。

11. Kitchen cooks should know what to do in case of a fire.
    厨师应该知道发生火情应该怎么做。

12. Kitchen cooks should know the fire department number.
    厨师应该知道消防部门的电话号码。

13. Kitchen cooks should understand related regulations.
    厨师应该了解相关规章制度。

14. Kitchen cooks should know how to use kitchen extinguishers.
    厨师应该知道如何使用厨房灭火器。

15. Kitchen cooks should understand cooking terms in English.
    厨师应该知道如何用英语来表达厨房术语。

16. Kitchen cooks should know how to handle knives safely.
    厨师应该知道如何安全使用刀具。

17. Kitchen cooks should know correct plate presentation and garnishment.
    厨师应该知道正确的装盘和装饰方法。

18. Kitchen cooks should present food in an appealing manner.
    厨师出品的菜肴样式应该新颖别致。

19. Kitchen cooks should meet even the most difficult requests with grace and courtesy.
    即使是最难满足的要求，厨师也应该心平气和，礼貌对待。

Unit 2

Fruit
水 果

You will be able to:

1. Remember the names of fruit in English.
2. Remember the useful verbs of preparing fruit.
3. Grasp the useful expressions in the dialogues.

Unit 2 Fruit 水果

Part A: Fruits

pineapple 菠萝

strawberry 草莓

blueberry 蓝莓

mango 杧果

cherry 樱桃

litchi 荔枝

orange 橙子

banana 香蕉

papaya 木瓜

watermelon 西瓜

apple 苹果

kiwi 猕猴桃

grape 葡萄

pear 梨

peach 桃

mangosteen 山竹

apricot 杏

pomegranate 石榴

pomelo 柚子

date 枣

Words & Expressions

strawberry ['strɔ:bəri] *n.* 草莓
mangosteen ['mæŋgəsti:n] *n.* 山竹
pomegranate ['pɔmigrænit] *n.* 石榴
papaya [pə'paiə] *n.* 木瓜
apricot ['eiprikɔt] *n.* 杏
pomelo ['pɔmiləu] *n.* 柚子

Talk about the shapes and flavors of the fruits following the example.

A: What's the shape of the orange?
B: It's round.
A: What's the flavor of it?
B: It's sweet and sour.

形容味道的词有：酸（sour）、甜（sweet）、苦（bitter）、辣（spicy）、咸（salty）。要记住哦！

Part B: English terms of processing fruit

put apples into the fruit salad	把苹果放入水果沙拉
wash the pears	洗梨
soak the strawberries	浸泡草莓
peel the oranges	给橘子剥皮
squeeze the lemons	榨柠檬汁
remove the seeds from the watermelon	把西瓜籽去掉
refrigerate the melon	把瓜（放进冰箱）冷藏
slice the watermelon	切开西瓜
mash up the bananas	把香蕉捣烂

Words & Expressions

process [prə'ses] v. 加工，处理   wash [wɔʃ] v. 清洗

peel [pi:l] v. 剥皮   soak [səuk] vt. 浸泡，浸透

squeeze [skwi:z] v. 榨，挤   remove [ri'mu:v] v. 去掉，搬开

refrigerate [ri'fridʒəreit] vt. 冷藏，冷冻   mash up 捣烂，压碎，压扁

Part C: Dialogues

Dialogue 1

Chef: Hi, Jack!

Jack: Yes?

Chef: Let's prepare a fruit salad!

Jack: Fruit salad? What fruit do we need?

Chef: We need one pineapple, three oranges, one watermelon and some strawberries.

Jack: OK. I'm going to prepare all of these right now!

Chef: Thank you.

Jack: My pleasure.

Words & Expressions

salad ['sæləd] n. 沙拉   prepare [pri'peə(r)] v. 准备

Notes:

1. Let's prepare a fruit salad. 让我们准备做水果沙拉。

fruit salad 水果沙拉

2. What fruit do we need? 我们需要哪种水果？

Dialogue 2

Jack: What should I do?

Chef: Give me a bunch of green grapes.

Jack: All right.

Chef: Wash them thoroughly.

Jack: It's not easy to make them clean.

Chef: Soak them in salted water for a while.

Jack: OK. The grapes are ready. What are you going to do with them?

Chef: I'm going to make fresh grape juice.

Words & Expressions

salted ['sɔːltid] *adj.* 盐腌的，盐味的

juice [dʒuːs] *n.* 果汁

fresh [freʃ] *adj.* 新鲜的

bunch [bʌntʃ] *n.* 束，串，捆

Notes:

1. Give me a bunch of green grapes. 给我一串绿葡萄。

   a bunch of 一串，一束，一群

   例如：a bunch of flowers 一束花

   　　　a bunch of people 一群人

2. Soak them in salted water for a while. 在盐水中浸泡一会儿。

Drills

Drill One

A: What do you want me to do?

B: Bring me
- a bunch of green grapes.
- some pears.
- one peach.
- two watermelons.

Drill Two

A: What are you going to do?

B: I'm going to
> make fresh grape juice.
> peel the oranges.
> make a fruit salad.
> squeeze the lemons.

Exercises

Ⅰ. Match the correct words.

1. apple   A. 杧果
2. strawberry   B. 葡萄
3. orange   C. 柑橘
4. lemon   D. 苹果
5. grape   E. 猕猴桃
6. watermelon   F. 草莓
7. kiwi   G. 樱桃
8. pear   H. 柠檬
9. cherry   I. 梨
10. mango   J. 西瓜

Ⅱ. Discuss the following questions with your partner.

1. What fruit do you like best?
2. In which season do we have watermelon?
3. Have you had mangoes before?
4. Which do you prefer, kiwi or grape?
5. How do you clean the strawberries?

Ⅲ. Fill in the blanks with the important verbs we have learnt.

1. _____ the orange.   2. _____ the seeds from the mangoes.
3. _____ the bananas.   4. _____ the melon.
5. _____ the strawberries.   6. _____ open the watermelon.
7. _____ the lemons.   8. _____ the grapes.

Ⅳ. Complete the following dialogue and read it aloud.

> like   juice   help   thirsty   drink   bring

Susan: Claire, I'm too tired to walk another step!

Unit 2 Fruit 水果

Claire: Me too. And I'm also very ____1____ . Do you want something to ____2____ ?
Susan: Yes.
Waiter: May I ____3____ you?
Claire: Yes. We'd like fresh ____4____ . What kind of juice do you have?
Waiter: We have orange, pineapple, watermelon, papaya and mango juice. Which one would you like?
Claire: I'd ____5____ pineapple juice. What about you, Susan?
Susan: Mango juice.
Waiter: OK. One pineapple juice and one mango juice. I'll ____6____ them to you right now.

V. Translate the following sentences into English.
1. 我要做水果沙拉。
2. 你需要哪种水果？
3. 你要喝点什么？
4. 我想要一杯菠萝汁。
5. 橙子又甜又酸。

Part D: Reading

How to Make a Fruit Salad

Fruits are good for our health. Most people like to eat fruit salads. Today, let's learn how to make a fruit salad. You can follow the directions below.

Step 1: Selecting the fruits

Remember that fruit salads are often creative: you decide which fruits and how much to use. Select whatever fruits are in season, what you like or what you have on hand. You can choose some tasty fruits such as strawberries, apples, bananas, grapes, watermelons, mangoes, oranges and peaches.

Step 2: Preparing the fruits

Wash all the fruits first, and then peel fruits with a tough skin, such as banana and mango. Cut them into bite-sized pieces. Remove the seeds from the fruit such as watermelon and oranges.

Step 3: Mixing the dressing and the fruits

Put the dressing and the fruits in a large mixing bowl and then mix them evenly.

Step 4: Enjoy

Words & Expressions

direction [də'rekʃn] *n.* 用法，说明
tough [tʌf] *adj.* 坚硬的
seed [si:d] *n.* 种子
dressing ['dresiŋ] *n.* 调味汁

creative [kri'eitiv] *adj.* 创造性的
bite-sized *adj.* 一口大小的
evenly ['i:vnli] *adv.* 均匀地

Unit 3

Vegetable
蔬 菜

You will be able to:

1. Remember the English words of vegetables.
2. Remember the useful verbs of processing vegetables.
3. Grasp the useful expressions in the dialogues.

烹饪厨房英语

 Part A: Vegetables

Words & Expressions

broccoli ['brɔkəli] *n.* 西兰花
lotus root ['ləutəs ru:t] *n.* 莲藕
scallion ['skæliən] *n.* 青葱，大葱
cucumber ['kju:kʌmbə（r）] *n.* 黄瓜

lettuce ['letis] *n.* 莴苣，生菜
spinach ['spinitʃ] *n.* 菠菜
asparagus [ə'spærəgəs] *n.* 芦笋

Unit 3 Vegetable 蔬菜

Talk about the colors of the vegetables following the example.

A: What's the color of carrot?
B: It's orange.
A: What's the color of eggplant?
B: It's purple.

形容颜色的词有：red orange yellow green blue purple white black pink等。

Part B: English terms of processing vegetables

wash the tomato	洗西红柿
peel the asparagus	芦笋去皮
slice the celery	芹菜切片
shred the onion	洋葱切丝
dice the carrot	胡萝卜切丁
cut the potato into slices/dices/shreds	把土豆切成片/丁/丝
mash the potato	捣土豆泥
boil the corn	煮玉米
slice the cauliflower	掰花菜
chop the onions fine	把洋葱切细点

Words & Expressions

process [prə'ses] v. 加工，处理
peel [pi:l] v. 剥皮
slice [slais] v. & n. 切成片；薄片
boil [bɔil] vt. 用开水煮
split up （把……）分成若干较小的部分

mash [mæʃ] v. 捣成泥
shred [ʃred] v. & n. 切丝；丝
dice [dais] v. & n. 切丁；骰子，方块
cut... into... 把……切成……形状
chop [tʃɔp] v. 砍，伐，劈

Part C: Dialogues

Dialogue 1

Chef: Could you do me a favor?
Jack: Yes, of course.
Chef: Help me process the potatoes, please.
Jack: OK. How can I do with the potatoes?
Chef: First, you should peel them, and then cut them into slices.
Jack: OK.

 烹饪厨房英语

Chef: Last, soak them into the cold water.
Jack: To prevent them becoming brown?
Chef: Yes.

Words & Expressions

do sb. a favor　帮某人一个忙　　　　　process [prəˈses] v.　加工，处理
soak [səuk] vt.　浸泡，浸透，吸入　　prevent [priˈvent] vt.　预防，阻碍，防止

Notes:

1. Could you do me a favor?　你能帮我一个忙吗？
2. Help me process the potatoes, please.　请帮我处理一下土豆。
   help sb. do sth　帮助某人做某事
3. first... then... last...　先……再……最后……（表示描述事情的先后顺序）
4. To prevent them becoming brown?　防止土豆变色吗？（这里的"them"指的是土豆）

Dialogue 2

Chef: Wash the green peppers.
Jack: OK.
Chef: Then cut the green peppers open.
Jack: In order to remove the seeds?
Chef: Yes. Stuff the peppers with meat.
Jack: How will we cook them?
Chef: We'll cook the green peppers in tomato sauce.
Jack: How about the red peppers?
Chef: Finely slice the red peppers.
Jack: To make salad?
Chef: Yes. For salad.

Words & Expressions

remove [riˈmu:v] v.　去掉，搬开　　　　seed [si:d] n.　种子，籽
stuff [stʌf] vt.　把……装满，填，塞　　salad [ˈsæləd] n.　沙拉

Unit 3 Vegetable 蔬菜

Notes:

1. OK = all right "对" "好" "行" 的意思（美口语）
2. green pepper 青椒，red pepper 红椒
3. Then cut the green peppers open. 然后把青椒切开。（在这里，"open" 是形容词，表示"开的"，在句子中做宾语补足语。）
4. remove the seeds 把籽拿出来
5. stuff...with... 用……把……装满　stuff the peppers with meat 用肉把青椒装满
6. tomato sauce 西红柿酱（汁）
7. "How about...?" "……怎么样？"（用于征求意见或问消息。）

Drills

Drill One

A: How can I do with the potatoes?

B: You can
- peel
- slice
- shred
- dice
- mash

the potatoes.

Drill Two

A: What can I do for you?

B: Help me wash the
- cucumbers.
- carrots.
- eggplants.
- potatoes.
- celeries.

Exercises

Ⅰ. Match the correct words.

1. cauliflower　　　　A. 菠菜
2. eggplant　　　　　B. 洋葱
3. green pepper　　　C. 黄瓜
4. spinach　　　　　 D. 韭菜
5. tomato　　　　　　E. 大蒜
6. onion　　　　　　 F. 蘑菇

烹饪厨房英语

7. cucumber G. 花菜
8. leek H. 青椒
9. garlic I. 茄子
10. mushroom J. 西红柿

Ⅱ. Can you name these vegetables?

1. Name a long, thin, orange vegetable that grows underground. It starts with a "C".
2. Name a purple vegetable that starts with an "E".
3. Name an orange vegetable that can be made into pie. It starts with a "P".
4. Name a crisp, green vegetable that has long stalks. It starts with a "C".
5. Name a green, leafy vegetable that tastes good in salads. It starts with an "L".

Ⅲ. Fill in the blanks with the important verbs we have learnt.

1. _____ the tomato
2. _____ the asparagus
3. _____ the celery
4. _____ the onion
5. _____ the carrot
6. _____ the potato into slices
7. _____ the potato
8. _____ the corn
9. _____ the cauliflower
10. _____ the onions fine

Ⅳ. Complete the following dialogue and read it aloud in pairs.

> open   sauce   salad   wash   remove   with   about

A: ___1___ the green peppers.
B: OK.
A: Then cut the green peppers ___2___.
B: To ___3___ the seeds?
A: Yes. Stuff the peppers ___4___ meat.
B: How will we cook them?
A: We'll cook the green peppers in tomato ___5___.
B: What ___6___ the red peppers?
A: Finely slice the red peppers.
B: For ___7___ ?
A: Yes. For salad.

20

Unit 3 Vegetable 蔬菜

V. Translate the following sentences into English.

1. 你能帮我个忙吗？
2. 请帮我处理一下土豆。
3. 先把胡萝卜切丁，再把洋葱切丝。
4. 把土豆浸泡在凉水中。
5. 你能把黄瓜榨汁吗？
6. 先把青椒切开，然后把籽拿出来。

Part D : Reading

Poisonous tomatoes

To the French, the tomato is the apple of love; to the Germans, it's the apple of paradise. To North Americans and the British, at least until the early 19th century, it was wolf peach, and some thought that tomatoes were poisonous.

Hundreds of years ago, Americans never ate tomatoes. They grew them in their gardens because tomato plants are so pretty. But they thought the vegetable was poisonous. They called tomatoes "poison apples". English doctors warned patients that tomatoes were poisonous and would bring death to anybody who ate one.

President Thomas Jefferson, however, knew that tomatoes were good to eat. He was a learned man. He had been to Paris, where he learned to love the taste of tomatoes. He grew many kinds of tomatoes in his garden. The President taught his cook a way, for a cream of tomato soup. This beautiful pink soup was served at the President's next dinner party. The guests thought the soup tasted really good. They never thought their President would serve his dinner guests poison apples.

Today, the tomato is one of the most commonly consumed vegetables in the world. Recent studies suggest that the beta carotene（胡萝卜素） and lycopene（茄红素） in tomatoes may help prevent cancer and heart disease. They're also a good source of vitamin A, vitamin C, potassium（钾） and iron（铁）.

Words & Expressions

poisonous ['pɔizənəs] *adj.* 有毒的
warn [wɔ:n] *v.* 警告
taste [teist] *n. & v.* 味道，品尝
cancer ['kænsə(r)] *n.* 癌症
vitamin ['vitəmin] *n.* 维生素

paradise ['pærədais] *n.* 天堂，伊甸园
patient ['peiʃnt] *n.* 病人
consume [kən'sju:m] *v.* 消耗
heart disease 心脏病
prevent [pri'vent] *v.* 预防，防止

Unit 4

Meat & Poultry
肉和禽类

You will be able to:
1. Remember the English words of meat & poultry.
2. Remember the useful verbs of processing meat.
3. Grasp the useful expressions about processing meat in the dialogues.

Unit 4 Meat & Poultry 肉和禽类

🍲 Part A: Meat & Poultry

Ⅰ. 常见肉类与家禽

Words & Expressions

mutton ['mʌtn] n. 羊肉
fat [fæt] n. 肥肉
spare [speə] adj. 备用的，多余的
bacon ['beikən] n. 培根
sausage ['sɔsidʒ] n. 香肠
breast [brest] n. 胸脯

lean [li:n] adj. 瘦的
streaky ['stri:ki] adj. 不均匀的
rib [rib] n. 肋骨
liver ['livə(r)] n. 肝脏
claw [klɔ:] n. 爪

23

烹饪厨房英语

Ⅱ. 细分牛肉、猪肉

Beef 牛肉

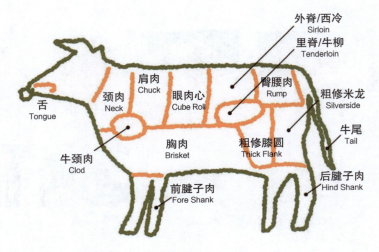

Words & Expressions

tongue [tʌŋ] *n.* 舌头　　　　　　　　brisket ['briskit] *n.* 胸肉
sirloin　西冷（外脊肉）　　　　　　shank [ʃæŋk] *n.* 小腿
rump [rʌmp] *n.* 臀部　　　　　　　silverside ['silvəsaid] *n.* 最上端股肉（粗修米龙）

Pork 猪肉

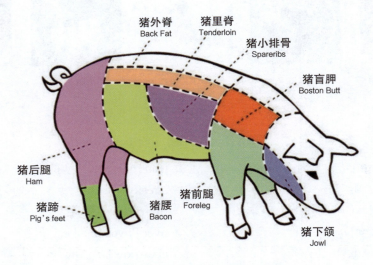

Words & Expressions

ham [hæm] *n.* 火腿　　　　　　　spareribs ['speəribz] *n.* 小排骨
jowl [dʒaul] *n.* 颌

Unit 4 Meat & Poultry 肉和禽类

🍲 Part B: English terms of processing meat

slice the meat	将肉切片
mince the meat	将肉切碎
mash the meat	将肉捣烂
dice the meat	将肉切丁
bone the meat	将肉去骨
skin the meat	将肉去皮

Words & Expressions

slice [slais] v. 切成薄片
mash [mæʃ] v. 把……捣成糊状
bone [bəun] v. 除去骨头

mince [mins] v. 切碎
dice [dais] v. 将……切成丁
skin [skin] v. 剥皮，削皮

🍲 Part C: Dialogues

Dialogue 1

Chef: Jack, could you wash the chicken breasts for me?
Jack: Sure.
Chef: Also, please mix the salt, flour and pepper.
Jack: I already did.
Chef: Good! Coat the chicken breasts evenly with the mixture.
Jack: It's done.
Chef: Now cook them!
Jack: Shall I cook them over low heat?
Chef: No. Cook over medium heat until light brown.

Words & Expressions

breast [brest] n. 胸部，胸脯
mixture ['mikstʃə(r)] n. 混合物，混合
light [lait] adj. 浅的，轻的

mix [miks] v. 使混合，混合
flour ['flauə(r)] n. 面粉
brown [braun] adj. 褐色的，棕色的

Notes:

1. Could you wash the chicken breasts for me? 你能帮我把这些鸡胸肉洗一下吗?
2. Also, please mix the salt, flour and pepper. 请再把那些盐、面粉和胡椒粉调一下。

25

烹饪厨房英语

mix sth. 混合某物

3. Coat the chicken breasts evenly with the mixture. 把这些调料均匀地涂在鸡胸肉上。

   coat... with... 给……涂上一层……

4. low heat 文火

Dialogue 2

Chef: Jack, what are you doing?

Jack: I am frying beef steak.

Chef: Mm, while cooking meat you should cook it according to the guests' preference.

Jack: Guests' preference?

Chef: Yes, some guests prefer their meat rare or medium rare. Some prefer medium, but some prefer medium-well or even well-done.

Jack: OK, thanks.

Chef: You are welcome, Jack.

Words & Expressions

customer ['kʌstəmə（r）] n.　顾客，主顾
prefer [pri'fə:（r）] v.　更喜欢
rare [reə（r）] adj.　三成熟的
medium ['mi:diəm] adj.　适中的，五成熟的

preference ['prefrəns] n.　喜爱，偏爱
would like　想要
well-done ['wel'dʌn] adj.　全熟的

Notes:

1. I am frying beef steak. 我正在煎牛排。

2. While cooking meat you should cook it according to the guests' preference. 做肉时，你要根据客人的喜好来做。

   according to 按照，根据

3. Some guests prefer their meat rare or medium rare. Some prefer medium, but some prefer medium-well or even well-done. 有些客人想要三成熟或四成熟。有些喜欢五成熟，而有些更喜欢七成熟甚至全熟。

Drills

Drill One

A: What shall I do with meat?

Unit 4 Meat & Poultry 肉和禽类

B: Cut it into slices.
Cut it into dices.
Cut it into shreds.
Cut it into segments.
Cut it into cubes.
Fry it in the wok.
Put it into the wok piece by piece.

Drill Two

A: Is | the meat / the beef / the chicken / the steak | cooked/done/finished?

B: It's done./No, it's not done yet.

Exercises

Ⅰ. Match the correct words.

1. pork   A. 瘦肉
2. roast duck   B. 火鸡
3. lean meat   C. 猪肉
4. spareribs   D. 牛舌
5. chicken breasts E. 鹅肝
6. smoked bacon   F. 烤鸭
7. beef steak   G. 肋排肉
8. goose liver   H. 鸡胸脯肉
9. turkey   I. 熏肉
10. tongue   J. 牛排

Ⅱ. Fill in the blanks with the words given below. And then read the sentences aloud.

out of   with   in   from   into

1. The cook is sticking a roasting fork _____ the meat.
2. The cook is pulling the roasting fork _____ the meat.

27

 烹饪厨房英语

3. The cook is taking the meat _____ the pan.

4. The cook is lifting the meat _____ the roasting fork.

Ⅲ. Fill in the blanks according to the Chinese meaning.

1. 将牛肉去骨。　　　　To bone the _____.
2. 在肉里塞入油脂。　　To lard the _____.
3. 装饰鸡肉。　　　　　To decorate the _____.
4. 将猪肉拍平。　　　　To flatten the _____.
5. 拍打五花肉。　　　　To beat the _____.

Ⅳ. Translate the following sentences into English.

1. 肉已经烤好了。
2. 主厨正在煎牛排。
3. 请再把那些盐、面粉和胡椒粉调一下。
4. 把这些调料均匀地涂在鸡胸肉上。
5. 做肉时，你要根据客人的喜好来做。

Ⅴ. Make short dialogues according to the following situations and do role-plays with your partners, using as many sentences as you can.

One person comes to your hotel in the afternoon. He/She is hungry and needs something to eat and drink. As a waiter / waitress, you need to take orders for him/her.

 Part D: Reading

Roast Duck: the Visiting Card of Beijing

Good morning everyone, my name is Yue Qingyang, I'm a girl of Beijing, and I'm 13 years old. I grow up with Beijing's water and food, and I armed my brains with brilliant cultures with Beijing characteristics. I love this city very much. I love to eat delicious food of Beijing, and I love to enjoy the eight beautiful views of Yanshan, and I love to appreciate the colorful cultures of Beijing. More important, I admire the persons who created all these excellent things very much. I have lots of things to say, but, today I only have time to introduce one of these things: Beijing roast duck.

The history of the roast duck can be traced back to as early as the Yuan Dynasty （1206-1368） when it was listed among the imperial dishes in the Complete Recipes for Dishes and Beverages, written in

1330 by Hu Sihui, an inspector of the imperial kitchen. Details regarding the cooking process were also described in this early cookbook.

The making process of the roast duck is complicated. The stuffed duck is hung in the roaster and kettles of hot water are placed in front to fill out the duck. Proper timing and temperature are important and the duck is turned often enough to roast the meat completely and evenly. (Try 350 °F turn every 15 minutes, total roasting time about 40 minutes). Roast until golden brown with rich grease perspiring outside and have a nice odor.

The last is how to eat it. There is a proper way to eat it. First, pick up a slice of duck with the help of a pair of chopsticks and dip it into the soy paste. Next, lay it on the top of a thin cake and add some bars of cucumber and shallot. Finally, wrap the stuff into a bundle with the sheet cake (a thin pancake). The real secret of roast duck's flavor lies in your carefully nibbling away at the mixture. You will find all the different ingredients very compatible. People say: "It's a pity to leave Beijing without trying the roast duck. The taste of the roast duck is in the eating."

I like to eat roast duck, do you like it?

Words & Expressions

brilliant ['briljənt] adj. 卓越的，灿烂的
appreciate [ə'priːʃieit] vt. 欣赏，领会
be traced back to 追溯到
recipe ['resəpi] n. 秘诀，食谱，药方
inspector [in'spektə(r)] n. 检查员，巡官
grease [griːs] n. 油脂
odor ['əudə] n. 气味，名声，气息
bundle ['bʌndl] n. 捆，束，包
compatible [kəm'pætəbl] adj. 一致的，兼容的，适合的

characteristic [ˌkærəktə'ristik] n. 特点，特性
admire [əd'maiə(r)] vt. 钦佩，赞美
imperial [im'piəriəl] adj. 皇帝的，皇家的
beverage ['bevəridʒ] n. 饮料
kettle ['ketl] n. 水壶，坑穴
perspire [pə'spaiə(r)] v. 出汗，分泌
wrap [ræp] v. 包，裹，覆盖
nibbling ['nibəliŋ] n. 一点一点地咬

Unit 5

Seafood
海 鲜

You will be able to:
1. Remember the English words of seafood.
2. Remember the useful verbs of prepare seafood.
3. Grasp the useful expressions in the dialogues.

Unit 5 Seafood 海鲜

Part A: Fish and Shellfish

Fish 鱼类

cod 鳕鱼

sea perch 海鲈鱼

salmon 三文鱼

trout 鳟鱼

tuna 金枪鱼

mandarin fish 鳜鱼

yellow croaker 黄花鱼

hairtail 带鱼

carp 鲤鱼

crucian carp 鲫鱼

grass carp 草鱼

big-head carp 大头鱼

mackerel 鲭鱼

sole 舌鳎

flounder 比目鱼

halibut 大比目鱼

sardine 沙丁鱼

wolf fish 海鲶鱼

shark 鲨鱼

grouper 石斑鱼

pomfret 鲳鱼

plaice 鲽鱼

eel 鳗鱼

skate 鳐鱼

Shellfish 贝类

abalone 鲍鱼	conch 海螺	clam 蛤蚌	oyster 牡蛎
scallop 扇贝	mussel 贻贝，淡菜	squid 鱿鱼	octopus 章鱼
cuttlefish 墨鱼	lobster 龙虾	crayfish 小龙虾	prawn 对虾
shrimp 小虾，河虾	crab 螃蟹	sea urchin 海胆	snail 田螺

Words & Expressions

cod [kɔd] n. 鳕鱼
salmon ['sæmən] n. 三文鱼
tuna ['tju:nə] n. 金枪鱼
yellow croaker ['krəukə] n. 黄花鱼
crucian carp ['kru:ʃən kɑ:p] n. 鲫鱼
flounder ['flaundə] n. 比目鱼
sardine [sɑ:'di:n] n. 沙丁鱼
pomfret ['pɔmfrit] n. 鲳鱼
abalone [æbə'ləuni] n. 鲍鱼
clam [klæm] n. 蛤蚌

sea perch [pə:tʃ] n. 海鲈鱼
trout [traut] n. 鳟鱼
mandarin fish ['mændərin fiʃ] n. 鳜鱼
hairtail ['heəteil] n. 带鱼
mackerel ['mækrəl] n. 鲭鱼
halibut ['hælibət] n. 大比目鱼
grouper ['gru:pə] n. 石斑鱼
plaice [pleis] n. 鲽鱼
conch [kɔntʃ] n. 海螺
oyster ['ɔistə] n. 牡蛎

Unit 5 Seafood 海鲜

scallop ['skɔləp] *n.* 扇贝
squid [skwid] *n.* 鱿鱼
cuttlefish ['kʌtlfiʃ] *n.* 墨鱼
crayfish ['kreifiʃ] *n.* 小龙虾
shrimp [ʃrimp] *n.* 小虾，河虾
sea urchin [si:'ə:tʃin] *n.* 海胆

mussel ['mʌsl] *n.* 贻贝，淡菜
octopus ['ɔktəpəs] *n.* 章鱼
lobster ['lɔbstə] *n.* 龙虾
prawn [prɔ:n] *n.* 对虾
crab [kræb] *n.* 螃蟹
snail [sneil] *n.* 田螺

Talk about your favorite seafood.

Waiter: Good evening, sir.
Guest: Good evening.
Waiter: May I help you?
Guest: I want to order some fish, would you like to make a recommendation?
Waiter: How about the mandarin fish?
Guest: Yes, I like it.
Waiter: How do you like your fish cooked, steamed or braised with brown sauce?
Guest: Just steamed.
Waiter: OK. We will serve it as soon as possible.

Words & Expressions

order ['ɔ:də] *vt.* 点餐，订购
steam [sti:m] *v.* 蒸

recommendation [,rekəmen'deiʃn] *n.* 推荐，介绍
braise [breiz] *vt.* 煨，焖

Part B: English terms of processing seafood

kill the eel — 杀鳗鱼
scale the cod and wash it clean — 鳕鱼去鳞、洗净
gut the mandarin fish — 鳜鱼去内脏
pack the fish fillet — 鱼排打包
bone the sardine — 沙丁鱼去骨
shell the conch — 海螺去壳
freeze the scallop — 扇贝保鲜冷藏
fry the squid quickly and slightly — 略微把鱿鱼爆炒一下
soak abalone in the warm water — 把鲍鱼放在温水中浸泡
slice sea cucumber diagonally into fillets and put it in a tray — 将海参切成斜刀片，放入碟中

33

 烹饪厨房英语

Words & Expressions

scale [skeil] vt. & n.　刮鳞；鱼鳞
pack [pæk] vt.　包装
bone [bəun] vt. & n.　去骨；骨头
shell [ʃel] vt. & n.　去壳；壳
slightly ['slaitli] adv.　轻微地
slice [slais] vt. & n.　切成片；薄片
tray [trei] n.　盘，碟

gut [gʌt] vt. & n.　取出内脏；内脏
fillet ['filit] vt. & n.　把（肉、鱼）切成片；肉片，鱼片
freeze [fri:z] vt.　用冷冻保藏（食物）
soak [səuk] vt.　浸，泡
diagonally [dai'ægənli] adv.　对角地，斜对地

Part C: Dialogues

Dialogue 1

Waiter: What would you like today?
Guest: I'm not sure.
Waiter: Shall I recommend the dishes for you?
Guest: Yes, of course.
Waiter: The beef is very good tonight. We also have lots of chicken dishes.
Guest: I'm not fond of chicken.
Waiter: What about fish? Fish is one of the specialities of the house. You can try stewed turtle, yellow croaker with sweet and sour sauce, braised mandarin fish with brown sauce, stir-fried shredded squid and steamed carp, etc.
Guest: I'll try yellow croaker with sweet and sour sauce.
Waiter: A good choice. Please wait a moment.

Words & Expressions

speciality [ˌspeʃi'æliti] n.　风味菜，特种菜品
turtle ['tə:tl] n.　海龟
stir-fry [stə: frai] v.　炒

stew [stju:] v.　炖，炖汤
sweet and sour sauce　糖醋汁
shred [ʃred] vt. & n.　裂为细条，撕成碎片；细条

Notes:

1. Fish is one of the specialities of the house. 服务员称自己所在餐馆一般用house代替restaurant。
2. be fond of　"喜欢……" 例如：I'm not fond of swimming.

Unit 5 Seafood 海鲜

Dialogue 2

Jack: What shall we do now?

Fish Cook: Process the fish.

Jack: Would you please teach me how to do with them?

Fish Cook: First, you should scale the fish.

Jack: With the fish scaler?

Fish Cook: Yes, maybe you can use a knife.

Jack: I know.

Fish Cook: To scale the fish, you should start from the tail towards the head. Then gut the fish and take away all the internal organs.

Jack: OK. Would you please show me now?

Fish Cook: Yes. Make a slit on the fish's belly, pull out the guts and rinse carefully to remove the blood.

Jack: OK. I will have a try.

Words & Expressions

process [prə'ses] vt. 加工
internal [in'tə:nl] adj. 内部的，在内部的
slit [slit] n. 狭长的切口，裂缝，裂口
rinse [rins] vt. 用清水洗

scaler ['skeilə] n. 刮鱼鳞器
organ ['ɔ:gən] n. 器官
belly ['beli] n. 肚子，腹部
remove [ri'mu:v] vt. 移动，搬开

Notes:

1. fish cook 鱼菜厨师
2. Would you please teach me how to do with them? 你能教我怎样处理鱼吗？

Drills

Drill One

A: What shall I do?

B: You should scale the
- salmon.
- yellow croaker.
- crucian carp.
- grass carp.

烹饪厨房英语

Drill Two

A: How can I do with the crucian carp?

B: You should | kill / scale / gut / slice / fry | the crucian carp.

Exercises

Ⅰ. Match the correct words.

Fish（鱼类）

1. sole   A. 鳟鱼
2. trout   B. 金枪鱼
3. tuna   C. 大比目鱼
4. shark   D. 鳜鱼
5. eel   E. 鲫鱼
6. cod   F. 带鱼
7. sardine   G. 三文鱼
8. salmon   H. 黄花鱼
9. mandarin fish   I. 鳗鱼
10. halibut   J. 沙丁鱼
11. mackerel   K. 鳕鱼
12. hairtail   L. 海鲈鱼
13. sea perch   M. 鲭鱼
14. yellow croaker N. 鲨鱼
15. crucian carp   O. 舌鳎

Shellfish（贝类）

1. prawn   A. 螃蟹
2. conch   B. 蛤蚌
3. lobster   C. 扇贝
4. crab   D. 章鱼
5. scallop   E. 鱿鱼
6. oyster   F. 龙虾
7. octopus   G. 牡蛎

Unit 5 Seafood 海鲜

8. clam   H. 鲍鱼
9. abalone   I. 海螺
10. squid   J. 对虾

II. Fill in the blanks with correct letters.

c_d   s_lmon   abal_ne   _yster   s_allop   c_nch   cl_m

shr_mp   l_bster   octop_s   c_ttlefish   pr_wn   s_uid   s_rdine

III. Fill in the blanks with the important verbs we have learnt.

1. _____ the sole   舌鳎去内脏
2. _____ the fish fillet   鱼排打包
3. _____ the cod   鳕鱼保鲜冷藏
4. _____ the yellow croaker   黄花鱼去骨
5. _____ the fish   杀鱼
6. _____ the clam   蛤蚌去壳
7. _____ the grass carp   草鱼去鳞
8. _____ the fish maw in warm water for a day   用温水泡鱼肚一天
9. _____ the sea perch diagonally into fillets   海鲈鱼切成斜刀片，放入碟中
   and put it in a tray
10. _____ the fish clean   鱼洗干净

IV. Complete the following dialogue and read it aloud.

> scale   gut   pour   fillet   turn on

Jack: What should I do with this fish?

Fish Cook: Take it easy. First ____1____ it, and then ____2____ it.

Jack: May I scale it with a knife?

Fish Cook: Yes, and gut it with the fish scissors.

Jack: I have washed the fish. What should I do next?

Fish Cook: ____3____ the fish.

Jack: OK.

Fish Cook: You must make sure there aren't any bones in the fillets.

Jack: I have done it already.

Fish Cook: ____4____ the sauce over the fish fillets. Put them in the pan, ____5____ the oven and prepare to stew them.

 烹饪厨房英语

Jack: For how long?
Fish Cook: For fifteen to twenty minutes.

Ⅴ. Make up a dialogue with the given conditions. Suppose one of you acts as the customer and the other as the waiter or the waitress. Write down the dialogue and practise reading it with your partner.

1. 一位客人进餐厅想找位子。
2. 服务员迎上去，安排座位，问客人想吃什么。
3. 客人点了糖醋鲤鱼、软炸里脊和一瓶啤酒。

Ⅵ. Translate the following sentences into English.

1. 我推荐几道菜给您好吗？
2. 鱼是本店的特色菜之一。
3. 这是菜单。您可以试试糖醋黄花鱼。
4. 你能教我怎样处理鱼吗？
5. 首先，你应该去掉鱼鳞。
6. 在鱼肚子这里开个口子，取出内脏，彻底冲洗并去除血迹。
7. 我们将在15分钟后上菜。
8. 您喜欢鱼怎样做？
9. 把带鱼放入油中略微爆炒一下。
10. 将鲤鱼切成斜刀片，然后放进盘子里。

 Part D: Reading

How to Buy Fresh Fish

To cook fish right the Chinese way, you must begin at the market, for the first thing is to tell your fish monger to keep the scales on. If the scales are removed, the skin will become dry and the taste will be wrong. If the fish is too long for the pot, you may have to cut it in two, but the whole fish looks much nicer. Always keep the head and the tongue in a fish cheek.

It is most important that fish should be very fresh. It goes bad very quickly and should be cooked as soon as possible. The following are criteria to look for when buying fresh fish:

• The gills should be bright red and moist.
• The eyes should be full and bright, not sunken and dull.
• The flesh should be firm, and not loose, and the tail straight, not drooping.
• There should be no unpleasant smell. A muddy odor does not mean that the fish is no longer

Unit 5 Seafood 海鲜

fresh, but rather that the fish was caught in muddy waters.

• The skin should be shiny, iridescent, tight and firmly attached to the flesh.

• The shiny, complete scales should stick firmly to the skin.

• The belly should not be swollen or faded.

Words & Expressions

criteria [krai'tiəriə] *n.*　判断的标准（criterion的复数形式）

sunken ['sʌŋkən] *adj.*　陷入（sink的过去分词）

flesh [fleʃ] *n.*　肌肉，肉

droop [dru:p] *v.*　低垂，下垂

iridescent [ˌiri'desnt] *adj.*　（正式用语）呈虹彩的，现晕光的

belly ['beli] *n.*　肚子，腹部

fade [feid] *v.*　褪色

gill [gil] *n.*　鳃

moist [mɔist] *adj.*　潮湿的，润湿的

dull [dʌl] *adj.*　不清楚的，不鲜明的

firm [fə:m] *adj.*　坚硬的，坚固的，坚实的

muddy ['mʌdi] *adj.*　多泥的

attach [ə'tætʃ] *v.*　附上，加上，贴上

swollen ['swəulən] *adj.*　膨胀的，肿起的（swell的过去分词）

Unit 6

Dairy & Egg Products
乳制品和蛋类制品

You will be able to:

1. Remember the English words of dairy and egg products.
2. Remember the useful verbs of processing the eggs.
3. Grasp the useful expressions in the dialogues.

Unit 6 Dairy & Egg Products 乳制品和蛋类制品

Part A: Dairy and Egg Products

Milk 牛奶

whole milk
全脂牛奶

low-fat milk
低脂牛奶

skim milk
脱脂牛奶

condensed milk
炼乳

Cream 奶油

regular cream
普通奶油

heavy cream
浓奶油

sour cream
酸奶油

yogurt
酸奶

butter
黄油

cheese
奶酪，芝士

Eggs 鸡蛋

Boiled Eggs 水煮蛋

soft-boiled eggs
水煮嫩蛋

hard-boiled eggs
水煮老蛋

poached eggs
荷包蛋，水波蛋

Fried Eggs 煎蛋

sunny-side up
单面煎蛋

over easy
双面煎蛋

scrambled eggs
炒鸡蛋

omelet
煎蛋卷

 烹饪厨房英语

Words & Expressions

skim [skim] *vt.* 自液体表面撇取（油脂、浮渣等） condense [kən'dens] *vt.* 压缩，浓缩
regular ['reɡjulə] *adj.* 普通的   sour [sauə] *adj.* 酸的
boil [bɔil] *vt.* 煮沸   poach [pəutʃ] *vt.* 水煮（荷包蛋）
scramble ['skræmbl] *vt.* 炒（蛋）   omelet ['ɔmlit] *n.* 煎蛋卷

Talk about the ways of processing the eggs.

A: Do you often have eggs?

B: Yes, almost every day.

A: How can you process them in the kitchen?

B: We can fry, boil, scramble or poach them.

Part B: English terms of processing eggs

fried eggs   煎鸡蛋
scramble eggs   炒鸡蛋
boil eggs   煮鸡蛋
poach eggs   水煮荷包蛋

Part C: Dialogues

Dialogue 1

Waiter: Can I take your order?

Guest: Yes, I'd like the eggs, a bowl of cereal and some toast.

Waiter: How would you like your eggs?

Guest: Fried.

Waiter: Sunny-side up or over easy?

Guest: Over easy.

Waiter: And, to drink?

Guest: A cup of tea.

Waiter: Anything else?

Guest: No, that's all.

Unit 6 Dairy & Egg Products 乳制品和蛋类制品

Words & Expressions

cereal ['siəriəl] *n.* 麦片，谷类
fried [fraid] *adj.* 煎的

toast [təust] *n.* 烤面包（片），吐司
else [els] *adj.* 其他的

Notes:

1. How would you like your eggs? 您的鸡蛋要怎样做？
2. Anything else? 您还要别的吗？
3. No, that's all. 不用，就这样。

Dialogue 2

Jack: What shall we do with these eggs?

Chef: Scramble them.

Jack: Scramble eggs!

Chef: Yes. Crack three eggs in a bowl.

Jack: OK. Crack three eggs.

Chef: Add a dash of milk, salt and pepper, and then mix them.

Jack: OK. It is done. Now, should I heat a skillet over a medium-low flame?

Chef: Right. Melt 1/2 tbsp of butter into the skillet and pour the beaten eggs into it.

Jack: Then I will stir them off the bottom of the skillet gently.

Chef: Good. Continue cooking until the eggs are the consistency you like.

Jack: OK.

Words & Expressions

crack [kræk] *vt.* 打裂，击裂
skillet ['skilit] *n.* 平底锅，煎蛋锅
flame [fleim] *n.* 火焰
pour [pɔ:] *vt.* 倒，灌，注
stir [stə:] *vt.* 搅拌，搅起
bottom ['bɔtəm] *n.* 底部
consistency [kən'sistənsi] *n.* 浓度，硬度

dash [dæʃ] *n.* 少许，少量
medium ['mi:djəm] *n.* 中间，适中
melt [melt] *vi.* 融化，熔化
beat [bi:t] *vt.* 打，搅拌（蛋、奶油等）使起泡（成糊），beaten是其过去分词
gently ['dʒentli] *adv.* 轻轻地

Notes:

1. a dash of 表示少许、少量的意思。
2. tbsp 是tablespoon 的缩写，"汤匙"的意思。
   1 tablespoon（汤匙）= 3 teaspoons（茶匙）

 烹饪厨房英语

Drills

Drill One

A: How would you like your egg?

B: I would like it
- boiled.
- scrambled.
- poached.
- over easy.
- sunny-side up.

Drill Two

A: Could you give me some
- condensed milk?
- whole milk?
- butter?
- cheese?
- sour milk?

B: All right.

Exercises

Ⅰ. Match the correct words.

1. bolied eggs	A. 蛋卷
2. fried eggs	B. 冰激凌
3. scrambled eggs	C. 炒鸡蛋
4. poached eggs	D. 黄油
5. omelets	E. 荷包蛋
6. milk	F. 炼乳
7. condensed milk	G. 奶酪
8. cream	H. 煎鸡蛋
9. butter	I. 浓奶油
10. whole milk	J. 牛奶
11. heavy cream	K. 煮鸡蛋
12. ice cream	L. 低脂牛奶
13. yogurt	M. 酸奶
14. cheese	N. 全脂牛奶
15. low-fat milk	O. 奶油

Unit 6 Dairy & Egg Products 乳制品和蛋类制品

II. Fill in the blanks with correct letters.

sk_m   omel_t   c_ndnse   yog_rt   p_ _ch

cre_m   scr_mble   b_tter   so_r

III. Complete the following dialogue and read it aloud.

> variety   taste   attractive   recommend   menu

Waitress: Are you ready to order, sir?

Jack: Yes. I would like some wine tonight. Can you ____1____ something for me?

Waitress: Our red wine is very famous. You can have a try.

Jack: All right. What kind of red wine do you serve?

Waitress: Here are a ____2____ of red wines for you to choose. You can have a ____3____ and then make your decision. Here is the wine ____4____.

Jack: The dry red wine looks quite good. I'd like to try this one.

Waitress: Then do you like any cheese to go with it?

Jack: Why? Do I have to order that?

Waitress: Oh, no. But cheese is the best thing to go with red wine. And we have really fresh cheese which will be free if you order red wine. Here are the pictures. Please have a look and pick out something you like.

Jack: That's really nice. I'd like this one. It looks very ____5____.

Waitress: OK. Please wait a moment.

IV. Fill in the blanks with the important verbs we have learnt.

1. _____ eggs   煎鸡蛋
2. _____ eggs   炒鸡蛋
3. _____ eggs   煮鸡蛋
4. _____ eggs   水煮荷包蛋

V. Translate the following sentences into English.

1. 您的鸡蛋要怎么做？
2. 请在碗里打两个鸡蛋。
3. 我要一份蛋卷，配一点番茄酱。
4. 我早餐要一杯脱脂牛奶、两片面包和一份熏肉。
5. 还要别的东西吗？

 烹饪厨房英语

Part D: Reading

How to Fry an Egg Sunny-side Up

Can you make the perfect sunny-side up egg? Now let's see how to make it.

Instructions:

1. Turn on the stove burner at medium high heat.

2. Grease the frying pan with cooking spray or with one tablespoon of vegetable oil, corn oil or butter.

3. Wait a few minutes for the grease to heat up. Probably heated cooking oil will seem to have a consistency similar to that of water. Heated butter will sizzle but will not yet be brown.

4. Crack the egg over the pan into the hot grease.

5. Turn the heat down to low.

6. Let the egg fry for a few minutes. It will be ready when the bottom of the egg starts to brown and top coagulates. Watch it closely as it cooks.

7. Remove the egg from the frying pan with a plastic slotted spatula. If the frying pan was greased enough, or if you are using a non-sticky pan, you should be able to slide the egg onto a plate.

Tips:

1. Cook the egg on low heat so that the bottom side of the egg doesn't burn before the top is cooked.

2. Be sure to watch it and not let it overcook. The egg won't be considered sunny-side up if the yolk is cooked all the way through.

Words & Expressions

grease [gri:s] *vt.* & *n.* 为……涂（或抹）油；油脂

spray [sprei] *n.* 喷雾器

sizzle ['sizl] *vi.* 发出"嘶嘶"声（如油炸食物时的声音）

coagulate [kəu'ægjuleit] *vi.* （指液体）凝结，凝固

slot [slɔt] *vt.* 在……上开长孔，开槽于

spatula ['spætjulə] *n.* 刮铲

sticky ['stiki] *adj.* 黏的，黏性的

slide ['slaid] *v.* （使）在光滑表面上滑动或滑行

Unit 7

Drinks
饮 品

You will be able to:

1. Remember the English words of drinks.
2. Describe the flavors of drinks.
3. Grasp the useful expressions about drinks in the dialogues.

烹饪厨房英语

Part A: Drinks

red wine
红葡萄酒

white wine
白葡萄酒

whisky
威士忌

champagne
香槟

beer
啤酒

samshu
烧酒

Sake
日本清酒

rice wine
米酒

mineral water
矿泉水

milk
牛奶

yoghurt
酸奶

coffee
咖啡

milk shake
奶昔

fruit juice
果汁

tea
茶

soda
苏打水

ginger ale
干姜水

7-UP
七喜

Coca Cola
可口可乐

Pepsi Cola
百事可乐

Words & Expressions

whisky ['wiski] n. 威士忌　　champagne [ʃæm'pein] n. 香槟
samshu ['sæmsju:] n. 烧酒　　Sake [sɑ:ki] n. 日本清酒
mineral ['minərəl] n. 矿物质　yoghurt ['jəugə:t] n. 酸奶
shake [ʃeik] n. 奶昔　　　　soda ['səudə] n. 苏打水
ale [eil] n. 麦芽酒

Talk about the flavors of the drinks following the example.

A: How about the wine?
B: It tastes *smooth*.
A: How about the Whisky?
B: It tastes *strong*.

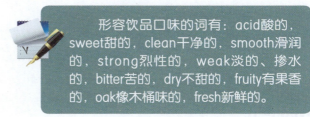

形容饮品口味的词有：acid酸的，sweet甜的，clean干净的，smooth滑润的，strong烈性的，weak淡的、掺水的，bitter苦的，dry不甜的，fruity有果香的，oak橡木桶味的，fresh新鲜的。

 ## Part B: English terms of drinks

mixed drink   调酒
with ice   加冰
without ice   不加冰

 ## Part C: Dialogues

Dialogue 1

A: What do you like to eat when you go out?
B: I like to eat pizza when I go out. When I'm home, I prefer to eat fish, rice, and soup.
C: Really? I like to eat noodles when I go out.
…
A: What do you like to drink with your meals?
B: It depends on the meal, but I usually like to drink water or tea. How about you?
C: I like to drink juice with breakfast, coffee with lunch, and water with dinner.

Words & Expressions

pizza ['piːtsə] *n.* 比萨　　　　prefer [pri'fə(r)] to 更喜欢，宁愿
rice [rais] *n.* 米饭　　　　　　soup [suːp] *n.* 汤
noodle ['nuːdl] *n.* 面条　　　　meal [miːl] *n.* 膳食
breakfast ['brekfəst] *n.* 早餐　　dinner ['dinə(r)] *n.* 晚餐，晚宴，主餐

Notes:

1. What do you like to eat when you go out? 你出去的时候喜欢吃什么？
   What do you like? 你喜欢什么？

 烹饪厨房英语

2. It depends on the meal, but I usually like to drink water or tea. 主要依饭而定，但是我通常喜欢喝白水或茶。

   depend on 依赖，依靠，取决于，依……而定

3. I like to drink juice with breakfast, coffee with lunch, and water with dinner. 我早饭喜欢喝果汁，午饭喜欢喝咖啡，晚饭喜欢喝白水。

Dialogue 2

A: May, what did you have for breakfast this morning?

B: I have some cakes and some juice. That's my favourite breakfast. What about you?

A: I have some bread, some milk and an apple. I like apples.

B: That sounds very healthy.

A: Yes, milk and apples are good for our health.

Words & Expressions

cake [keik] n. 蛋糕　　　　　favourite ['feivərit] n. 最喜爱的
milk [milk] n. 牛奶　　　　　healthy ['helθi] adj. 健康的
health [helθ] n. 健康

Notes:

1. May, what did you have for breakfast this morning? 梅，你今天早晨吃什么了呀？
   have=eat 吃

2. Milk and apples are good for our health. 牛奶和苹果对身体有好处。
   be good for 对……有好处

Drills

Drill One

A: Anything to drink?

B: Coffee, Grape juice, Tea, Soda, please.

Drill Two

A: What drink would you like?

B: I'll try
- whisky.
- red wine.
- Pepsi Cola.
- milk shake.

Exercises

Ⅰ. Fill in the blanks with correct letters.

香槟 ch_mp_gne 烧酒 sa_sh_
酸奶 _ogh_rt 果汁 j_ice
可口可乐 Coc_ Col_ 茶 t_ _

Ⅱ. Match the correct words.

1. white wine   A. 牛奶
2. grape juice   B. 白葡萄酒
3. mineral water   C. 啤酒
4. milk   D. 葡萄汁
5. Sake   E. 酸奶
6. sodas   F. 苏打水
7. champagne   G. 矿泉水
8. beers   H. 香槟酒
9. yoghurt   I. 日本清酒
10. coffee   J. 咖啡

Ⅲ. Fill in the blanks with the words given below. And then read the passage aloud.

how many, have had, too much, make, what, drink, of, should, other, sweet

Today I'd like to tell you something about chocolates. Most of us _____1_____ chocolates, but do you know _____2_____ states of the chocolates there are and _____3_____ they are made of?

Chocolate has three states. People _____4_____ the first state of chocolate from the seeds（种子）_____5_____ cacao trees. The second is a kind of _____6_____. It is made by mixing the first state of chocolate with hot water or milk. And the third is a kind of _____7_____ thing. It is made of the first state of chocolate, sugar and some _____8_____ delicious things. In China, a lot of people like to eat chocolates as sweets. But some doctors say people _____9_____ not have too many

 烹饪厨房英语

chocolates because they have _____10_____ fat and sugar.

Ⅳ. Translate the following sentences into English.
1. 我最喜欢的饮品是威士忌。
2. 请给我一杯红葡萄酒。
3. 柠檬汁喝起来有点酸。
4. 中国的茶很有名。
5. 牛奶和苹果对身体有好处。
6. 中国的白酒喝起来很浓烈。
7. 我早饭喜欢喝果汁，午饭喜欢喝咖啡，晚饭喜欢喝白水。
8. 主要依饭而定，但是我通常喜欢喝白水或茶。
9. 我妈妈喜欢喝咖啡。
10. 喝太多酒对我们身体不好。

Ⅴ. Pair Work.
Make a dialogue with your partner about ordering drinks in a restaurant. Then role-play the dialogue.

Part D: Reading

The Origins of Afternoon Tea

In 1662, King Charles Ⅱ of England married the Portuguese princess Catherine of Braganza who brought with her, as part of her dowry, a small chest of tea. As the new queen, Catherine began the serving of tea to her friends at court.①

Although there is mention of "five o'clock tea" in France in the 17th century, the credit for the invention of "Afternoon Tea" is given to Anna Russell, duchess of Bedford, during the long gap between an early breakfast and very late dinner, experienced what she called "a sinking feeling" ② at about 4 or 5 in the afternoon. She asked her maid to bring her a pot of tea, a little bread and butter and cake in her room. She found this arrangement so agreeable that she began asking her friends to join her.

High tea, a term often confused with afternoon tea, usually takes the place of supper.③ During the Industrial Revolution, working class families would return home tired and exhausted. The table would be set with dinner foods like meat, bread, butter, potatoes, cheese and of course tea. It was termed "high tea" because it was

eaten at a high dining table rather than a low tea table.

 Afternoon tea (because it was usually taken in the late afternoon) is also called "low tea" as it is served at a low table. Since this wasn't a meal, but rather like a late afternoon snack meant to stave off hunger, finger foods were the common fare. Tiny, dainty sandwiches, scones and pastries were served with afternoon tea.④ Finger foods afforded the opportunity to take a petite bite and easily maintain a conversation. This is most important, as one is not merely taking tea to gain nourishment or satisfy hunger, but to take time to relax, converse and enjoy the company of dear friends. In England, the traditional time for tea was four o'clock or five o'clock in the afternoon and no one stayed after seven o'clock.

Words & Expressions

dowry ['dauri] *n.* 嫁妆
court [kɔ:t] *n.* 宫廷
arrangement [ə'reindʒmənt] *n.* 布置，安排
confuse [kən'fju:z] with 弄混，混淆
term [tə:m] *vt.* 把……称为
stave [steiv] off 延缓，赶走
fare [feə(r)] *n.* 食物
scone [skɔn] *n.* （英国的一种圆饼）烤饼，司康饼
petite [pə'ti:t] *adj.* 娇小的
nourishment ['nʌriʃmənt] *n.* 营养

chest [tʃest] *n.* 箱子
credit ['kredit] *n.* 荣誉
agreeable [ə'gri:əbl] *adj.* 愉快的
exhausted [ig'zɔ:stid] *adj.* 精疲力竭的
snack [snæk] *n.* 小吃，点心
finger foods 手指食物（便于用手指取食的食物）
dainty ['deinti] *adj.* 小巧精致的，可口的
afford [ə'fɔ:d] *v.* 提供
maintain [mein'tein] *vt.* 维持
stay [stei] *v.* 停留，逗留

Notes:

1. 1662年，英国国王查理二世迎娶了葡萄牙布拉干萨王朝的凯瑟琳公主。公主随嫁妆带来了一小箱茶叶。当上了皇后的凯瑟琳开始在宫廷中以茶待客。
2. 一种虚脱感。
3. 人们常常会把"茶点"和"下午茶"搞混，其实茶点通常相当于晚餐。
4. 美味的小块三明治、烤饼和酥皮糕点通常都是享用下午茶时的小吃。

Unit 8

Dried Foods
干 货

You will be able to:

1. Remember the English words of dried foods.
2. Remember the useful terms about dried foods.
3. Grasp the useful expressions about doing shopping in the grocery.

Unit 8 Dried Foods 干货

Part A: Dried Foods

mushroom 蘑菇
sunflower seed 葵花子
dried fungus 干木耳
preserved fruit 蜜饯

dried day lily 干黄花菜
gingko 银杏果，白果
dried date 干枣
raisin 葡萄干

dried longan 干龙眼，干桂圆
almond 杏仁
walnut 核桃
cashew 腰果

chestnut 板栗
pinenut 松子，松仁儿
peanut 花生
hazelnut 榛子

Words & Expressions

fungus ['fʌŋgəs] n.　木耳
gingko ['giŋkəu] n.　银杏果，白果
longan ['lɔŋgən] n.　龙眼，桂圆
chestnut ['tʃesnʌt] n.　栗树，板栗
hazelnut ['heizlnʌt] n.　榛子

preserved [pri'zə:vd] adj.　防腐的，可保存的
raisin ['reizn] n.　葡萄干
almond ['ɑ:mənd] n.　杏树，杏仁
pinenut ['painʌt] n.　松子，松仁儿
cashew ['kæʃu:] n.　腰果

Part B: English terms of some dried foods

black fungus　　　　　黑木耳
white fungus　　　　　银耳
preserved fruits　　　　蜜饯

 烹饪厨房英语

black raisins	黑葡萄干
dried longan	干龙眼
almond juice	杏仁汁
chestnut pies	板栗饼
pinenut candies	松子糖
hazelnut chocolates	榛子巧克力

Part C: Dialogues

Dialogue 1

Mary: I'm home, Mum! What are you doing there?

Mother: I'm going shopping, and I am making a shopping list now, Mary.

Mary: Let me give you a hand.

Mother: All right, Mary. Are there any dried dates, funguses or honey in the cupboard?

Mary: Let me have a look, Mum.

  Oh, there aren't any dates and funguses, but there is some honey.

Mother: Are there any walnuts or peanuts in the box?

Mary: There aren't any of them.

Mother: Oh, dates, fungus, walnuts, pinenuts…

Mary: I want to have some bananas, Mum.

Mother: OK. I'll buy some for you. Thank you for your help, Mary.

Mary: I'm glad to be of your help.

Words & Expressions

shopping list　购物单　　　　　　　　　　honey ['hʌni] n.　蜂蜜
cupboard ['kʌbəd] n.　柜橱，食物柜，食橱，密室

Notes:

1. Let me give you a hand. 让我帮你一把。
   give sb. a hand 帮助某人
2. Let me have a look, Mum. 妈妈，让我看看。
   have a look 看
3. I'm glad to be of your help. 我很高兴能够帮助你。

Unit 8 Dried Foods 干货

Dialogue 2

A: Good evening, madam.

B: Good evening.

A: Take your seat, please! What would you like to have, madam?

B: Well, I'll begin with some mushroom soup and follow by some Chinese dishes. What do you recommend?

A: Well, black fungus is very famous in the north, would you like to try?

B: OK.

A: What would you like to drink?

B: A cup of water with honey, please.

A: All right. We will serve it 15 minutes later. And you can have some dried dates, almonds and some peanuts while waiting.

B: Thank you.

A: It's my pleasure.

Words & Expressions

soup [su:p] *n.* 汤
follow ['fɔləu] *vt.* 跟随
serve [sə:v] *vt.* 服务，上菜
recommend [ˌrekə'mend] *vt.* 推荐
north [nɔ:θ] *n.* 北部，北方

Notes:

1. I'll begin with some mushroom soup and follow by some Chinese dishes.
   我先喝点蘑菇汤，再吃点中国菜。
   begin with... 从……开始

2. What do you recommend ? 你推荐什么菜呢？

3. We will serve it 15 minutes later. 我们15分钟后上菜。

Drills

Drill One

A: I am making a shopping list now, are there any ____ in the cupboard?

| dried dates |
| black funguses |
| chestnuts |
| peanuts |

B: Let me have a look.

 烹饪厨房英语

Drill Two

A: What would you like to have, madam?

B: Well, I'll begin with some
- mushroom soup.
- seafood soup.
- snacks.
- fruit and vegetable salad.

Exercises

Ⅰ. Match the correct words.

1. preserved fruit   A. 花生
2. gingko   B. 干木耳
3. pinenut   C. 榛子
4. peanut   D. 葡萄干
5. walnut   E. 葵花子
6. mushroom   F. 蜜饯
7. dried fungus   G. 银杏果
8. raisin   H. 核桃
9. sunflower seed   I. 松子
10. hazelnut   J. 蘑菇

Ⅱ. Choose the best word or expression for each of the sentences.

1. You are told to cut the onions into _____.
   A. four   B. fourth   C. fourths
2. We need _____ cups of lemon juice.
   A. two and a half   B. two halves   C. two and half
3. I _____ brought the fruits.
   A. almost   B. always   C. already
4. How many _____ of olive oil shall I put?
   A. cloves   B. cups   C. tablespoons
5. You'd better cover the liquidizer, and blend it _____ high speed.
   A. in   B. at   C. with

Ⅲ. Complete the following dialogue and read it aloud.

> menu   recommend   else   fungus   reservation   delicious

Unit 8 Dried Foods 干货

A: Welcome to our restaurant! Do you have a ___1___ ?

B: Yes, a table for five.

A: Let me see. Yes, this way, please.

B: Thank you.

A: This is the ___2___ . May I have your order now?

B: Today we'd like to try some Chinese dishes. What do you ___3___ ?

A: Then I'll recommend you to try sauteed black ___4___ with sliced pork and eggs . It's our speciality .

B: Oh. What ___5___ can you recommend?

A: Well, we also have some mushroom soup as appetizers. It's very ___6___ .

B: OK. We want to try the dishes.

A: Just wait for a moment, please.

IV. Translate the following sentences into English.

1. 我在列一张购物单。
2. 壁橱里有栗子吗？
3. 让我看一看。
4. 我先喝点蘑菇汤。
5. 你推荐什么菜呢？

V. Role-play.

Your mother is making a shopping list. Use the words we have learnt in this unit. Make a dialogue between your mother and you.

Part D: Reading

Cooking Safely in Microwave Ovens

Microwave ovens can play an important role at mealtime, but special care must be taken when cooking or reheating meat, poultry, fish and eggs to make sure they are prepared safely.

Microwave ovens can cook unevenly and leave "cold spots", where harmful bacteria can survive. For this reason, it is important to use the following safe microwaving tip to prevent food borne illness.

Arrange food items evenly in a covered dish with a lid or plastic wrap; loosen or vent the lid or wrap to let steam escape. The moist heat that is created will help destroy harmful bacteria and ensure uniform cooking. Cooking bags also provide safe, even cooking.

Words & Expressions

microwave ['maikrəweiv] *n. & vt.* 微波，微波炉；用微波炉加热（或烹饪）
survive [sə'vaiv] *vi.* 幸存，活下来
mealtime ['mi:ltaim] *n.* 进餐时间
moist [mɔist] *adj.* 潮湿的，微湿的
food borne illness [医] 食物传染疾病
bacteria [bæk'tiəriə] *n.* 细菌
reheat [,ri:'hi:t] *vt.* 把……再加热
loosen ['lu:sn] *vt.* 解开或使松，放松
play an important role 扮演重要角色

Unit 9

Cooking Methods
烹饪方法

You will be able to:
1. Remember the English words of cooking methods.
2. Grasp the useful expressions in the dialogues.

烹饪厨房英语

Part A: Cooking methods

Words & Expressions

stir-fry ['stə:ˌfrai] v. 用旺火煸，用旺火炒
saute ['səutei] adj. & v. [法]炒的，嫩煎的；炒
braise [breiz] v. 炖，焖
grill ['gril] v. 烧烤，铁扒
poach ['pəutʃ] v. 水煮（荷包蛋）

roast ['rəust] v. 烤，烘，焙
stew ['stju:] v. 炖，煨

Part B: English terms of cooking methods

boil the corn   煮玉米
steam the rice   蒸米饭
fry beef steak   煎牛排

Unit 9 Cooking Methods 烹饪方法

braise the beef	炖牛肉
bake the bread	烤面包
deep-fry chicken wings	炸鸡翅
dip-boil mutton	涮羊肉
poach eggs	煮荷包蛋
roast the leg of lamb	烤羊腿

Words & Expressions

corn [kɔːn] *n.* 玉米
steak [steik] *n.* 牛排，肉排
lamb [læm] *n.* 羔羊，小羊；羔羊肉

rice [rais] *n.* 稻，稻米；大米
wing [wiŋ] *n.* 翅膀，翼

 Part C: Dialogues

Dialogue 1

Jack: What should I do?

Chef: Cook the onions.

Jack: OK.

Chef: Then mix the sesame seeds, water, garlic, salt, lemon juice and red peppers.

Jack: How much is the lemon juice?

Chef: Ten tablespoons.

Jack: Next?

Chef: Sprinkle the baking dish with breadcrumbs and parsley.

Jack: And then put the fish in the baking dish?

Chef: Yes. Pour the sesame seeds and onions over the fish.

Jack: Shall I cover the fish?

Chef: No. Don't cover the fish.

Jack: OK.

Chef: Did you light the oven?

Jack: No, not yet.

Chef: Light the oven, cook the fish at four hundred degrees.

Jack: For how long?

Chef: For twenty to twenty-five minutes.

Jack: I will garnish the fish with parsley and olives.

烹饪厨房英语

Words & Expressions

bake [beik] vt. 烤，烘焙
light [lait] vi. 点燃，发光，发亮
tablespoon ['teiblspu:n] n. 大汤匙，大调羹
breadcrumbs ['bredkrʌmz] n. 面包屑
olive ['ɔliv] n. 橄榄
sprinkle ['spriŋkl] vt. 洒，散置，把液体或颗粒洒在……上
garnish ['gɑ:niʃ] vt. 装饰，文饰，给……加配菜

baking dish 烤盘
oven ['ʌvn] n. 烤箱
sesame seed ['sesəmi][si:d] n. 芝麻籽
cover ['kʌvə(r)] n. 盖子，覆盖物
parsley ['pɑ:sli] n. [植] 西芹，欧芹

Notes:

1. How much...?   ……多少？（只用于不可数名词）
2. sprinkle... with...   把……撒在……上
3. pour... over...   把……倒在……上
4. at four hundred degrees   在这里at表示"（温度）达，以（……温度）"
5. For how long?   "for"在这里表示时间，即多久
6. garnish... with...   给……配上……

Dialogue 2

Jack: Chef, could you tell me the difference between basting and braising?

Chef: Basting and braising?

Jack: Yes.

Chef: Basting is a cooking method, which is using butter, sauce, or juice over to keep food moist during cooking.

Jack: It's to keep food moist.

Chef: Yes. Braising is searing food in oil and then simmering it.

Jack: So, basting is to keep the food moist, and braising is to simmer the food.

Chef: Right.

Jack: Thanks a lot!

Words & Expressions

difference ['difrəns] n. 差别，差异
butter ['bʌtə(r)] n. 黄油
sear [siə(r)] vt. 烧焦

baste [beist] v. 浇汁
moist [mɔist] adj. 潮湿的
simmer ['simə(r)] vt. 炖

Unit 9 Cooking Methods 烹饪方法

Notes:

1. Could you tell me the difference between basting and braising? 你能告诉我"浇汁"和"焖"在做法上有什么不同吗？
2. keep food moist 保持食物水分。

Drills

Drill One

A: What should I do?

B:
> Cook the onions.
> Mix the sesame seeds, water, garlic, salt and lemon juice.
> Sprinkle the baking dish with breadcrumbs and parsley.
> Put the fish in the baking dish.
> Pour the sesame seeds and onions over the fish.

Drill Two

A: What are you going to do?

B: I'm going to
> steam rice.
> boil noodles.
> roast chicken.
> stir-fry shredded potatoes.

Part D: Return to Pot Salmon

Ingredients

Salmon	250 g	Fermented soybeans	10 g
Garlic leaves	15 g	Vinegar	3 g
Green peppers	10 g	Chili oil	3 g
Red peppers	10 g	Green onions	5 g
Sugar	30 g	Pepper powder	3 g

烹饪厨房英语

Bean sauce	10 g	Starch	10 g
Egg	1 g	Salt	10 g
Garlic	2 g	Ginger	3 g

1. *Cut* the Salmon *into* slices. *Combine* the salt, pepper powder, starch, egg. *Mix* well and *reserve* for 10 minutes; cut the green & red peppers into cubes, cut the garlic leaves into strips and finely *chop* the green onions, ginger and garlic.

2. *Add* oil in a heated wok, *deep fry* the Salmon until deep brown in color and *remove*.

3. *Add* some oil in a heated wok, add fermented soybeans, *fry* until aromatic. Add bean sauce, green onions, ginger, garlic, a little water; add salt, sugar, vinegar, salmon and other ingredients. Fry, finish with the chili oil and remove.

Words & Expressions

wok [wɔk] *n.* 锅，炒菜锅
deep fry *v.* 炸
combine [kəm'bain] *v.* 使混合
ingredient [in'gri:diənt] *n.* （混合物的）组成部分，（烹调的）原料
strip [strip] *n.* 长条，条板
cube [kju:b] *n.* 立方形，立方体

add [æd] *v.* 加入
aromatic [ˌærə'mætik] *adj.* 有香味的
remove [ri'mu:v] *v.* 拿走，移动
slice [slais] *vt. & n.* 切成片；薄片

Exercises

Ⅰ. Match the correct words below.

1. roast   A. 煎
2. stir-fry   B. 铁扒
3. stew   C. 蒸
4. boil   D. 烤
5. grill   E. 红烧
6. steam   F. 焖
7. fry   G. 炖
8. red-cook   H. 爆炒
9. smoke   I. 煮
10. braise   J. 熏

Unit 9 Cooking Methods 烹饪方法

II. There are many different ways of preparing and cooking food. Take mushrooms for example, they can be stir-fried mushrooms, deep-fried mushrooms or grilled mushrooms. Now work in pairs and use the cooking methods and the foods given below.

Ways of cooking food:
A. frying
B. boiling
C. grilling
D. roasting
E. poaching
F. deep-frying
G. baking

Food for cooking:
1. duck
2. T-bone steak
3. bread
4. eggs
5. potatoes
6. chicken
7. salmon
8. fish

III. Complete the following dialogue.

| roast   heat   oven   roasting tray   put   call |

Jack: What do you _____ this tray?
Chef: It's a _____.
Jack: What are you going to do with it?
Chef: I'm going to _____ oil in it.
Jack: And then?
Chef: We'll _____ chicken.
Jack: I see. And what shall I do?
Chef: _____ the roasting tray in the _____.

IV. Translate the following sentences.
1. 在高压锅里蒸米饭。
2. 我今晚要吃涮羊肉。
3. 厨师正在做烤鸭。
4. 将鸡肉蒸一个半小时。
5. 把烤盘放在烤箱里。
6. Slice the mushrooms.
7. Peel the onions.
8. Steam the rice.

9. Fry the chicken.

10. Add some water.

Ⅴ. Rearrange the statements into the proper steps for preparing and cooking the dishes.

1. Cooking green peppers

   A. Cut the green peppers open.

   B. Cook the green peppers in tomato sauce.

   C. Stuff the green peppers with meat.

   D. Remove the seeds.

   E. Wash the green peppers.

2. Preparing carrots

   A. Boil the carrots in salt water.

   B. Cut the carrots into cubes.

   C. Wash the carrots carefully.

   D. Serve the carrots with butter.

   E. Peel the carrots.

Part E: Reading

Braised Dongpo Pork

Braised Dongpo Pork is a famous Hangzhou dish which is very popular in south of China. Legend has it that while Su Dongpo（a writer, poet, artist and statesman of the Song Dynasty）  was banished to Hangzhou in a life of poverty, he made an improvement of the traditional cooking process. He first braised the pork, added Chinese fermented wine and made braised pork in brown sauce, and then slowly stewed it on a low heat fire. This dish was first launched in Hangzhou, the capital of the Southern Song Dynasty, then flourished and became one of the most famous dishes in China.

Words & Expressions

legend ['ledʒənd] n. 传说，传奇人物

poverty ['pɔvəti] n. 贫穷，缺乏

dynasty ['dainəsti] n. 王朝，朝代

Braised Dongpo Pork 东坡肉

banish ['bæniʃ] vt. 放逐，驱逐

process [prə'ses] n. 过程，工序

flourish ['flʌriʃ] vi. 挥舞，茂盛，繁荣

Unit 10

Kitchen Equipment and Tools
厨房设备和厨房工具

You will be able to:
1. Remember the English words of kitchen equipment and tools.
2. Remember the English terms of cooking steps.
3. Grasp the useful expressions in the dialogues.

 Part A: Kitchen Equipment and Tools

Kitchen Equipment　厨房设备

oven　　　　　refrigerator　　　cupboard　　　stove
烤箱　　　　　冰箱　　　　　橱柜　　　　　煤气灶

steamer　　　　wok　　　　　frying pan　　　pressure cooker
蒸锅　　　　　炒锅　　　　　平底锅　　　　高压锅

microwave　　　casserole　　　grill　　　　　blender
微波炉　　　　砂锅　　　　　烤架　　　　　搅拌机

egg boiler　　　dish washer　　sink　　　　　mincer
煮蛋器　　　　洗碗机　　　　水池　　　　　绞肉机

Kitchen Tools　厨房工具

frying basket　　rolling pin　　　colander　　　scale
炸篮　　　　　擀面杖　　　　滤锅　　　　　秤

whisk　　　　　peeler　　　　　garlic press　　cutting board
搅拌器　　　　削皮器　　　　压蒜器　　　　菜板

Unit 10 Kitchen Equipment and Tools 厨房设备和厨房工具

cleaver
砍肉刀

paring knife
削皮刀

ladle
长柄勺

bottle opener
开瓶器

Words & Expressions

frying basket　炸篮
colander ['kʌləndə (r)] n.　滤锅，漏勺
whisk [wisk] n.　搅拌器
peeler ['pi:lə (r)] n.　去皮器，削皮器
cleaver ['kli:və (r)] n.　砍肉刀，剁肉刀
paring knife ['pεəriŋ naif] n.　削皮刀，去皮刀

rolling pin ['rəuliŋ pin] n.　擀面杖
scale [skeil] n. & vt.　秤；规模，等级；测量，刮去……的鳞片
press [pres] vt.& n.　压，按；逼迫；压平机；压榨机
ladle ['leidl] n.　长柄勺

Part B: English terms of cooking steps.

flatten the dough　　　　　　　　　　把面团擀平
mince the beef　　　　　　　　　　　把牛肉搅碎
blend at high/medium/low speed　　以高速/中速/低速搅拌
beat the eggs　　　　　　　　　　　把鸡蛋打散
crush the garlic　　　　　　　　　　把大蒜捣碎
scale the fish　　　　　　　　　　　把鱼鳞刮掉
heat the cooking oil in the frying pan　把油锅里的油加热
put the potato chips in the frying basket　把薯片放进油炸篮里
pour the ingredients into the baking pan　把所有材料倒进烤盘里

Words & Expressions

flatten ['flætn] v.　变平，使（某物）变平
blend [blend] vt.　混合，把……掺在一起
crush [krʌʃ] vt.　压破，压碎
pour [pɔ:(r)] vt.　涌出；倾，倒
ingredient [in'gri:diənt] n.　（烹调的）原料
potato chips n.　马铃薯片，炸土豆片

dough [dəu] n.　生面团
beat [bi:t] v.　接续击打
heat [hi:t] v.　使温暖，使热

烹饪厨房英语

Part C: Dialogues

Dialogue 1

Jack: What do you call these knives?

Chef: Well, this is a boning knife. A boning knife is used to remove bones from meat.

Jack: What is this funny little knife for?

Chef: It's an oyster knife. It's for opening oyster shells.

Jack: What are oysters? Can you tell me, chef?

Chef: They are a kind of shellfish living in the sea.

Jack: What do you want me to do now?

Chef: Carve the roast. OK?

Jack: Certainly. What kind of knife shall I use?

Chef: Use a carving knife. Look! It's over there.

Jack: Thank you very much, chef.

Words & Expressions

bone [bəun] *n. & vt.* 骨头；剔骨头　　　　oyster ['ɔistə (r)] *n.* 牡蛎

shell [ʃel] *n. & vt.* （贝、卵、坚果等的）壳；去壳　　shellfish ['ʃelfiʃ] *n.* 贝类动物

funny ['fʌni] *adj.* 有趣的，滑稽的

carve [kɑ:v] *vi. & vt.* 切，雕刻；切片，雕塑

Notes:

1. boning knife 去骨刀

2. oyster knife 开牡蛎刀

3. carving knife 雕刻刀

4. A boning knife is used to remove bones from meat.
   去骨刀是用来把肉从骨头上剔下来的。
   remove... from... 从……中移开，拿走，除掉
   例如：Please remove the seeds from the green pepper. 请把青椒籽去掉。

Dialogue 2

Jack: What do you call this tray?

Chef: It's a roasting tray.

Jack: What are you going to do with it?

Chef: I'm going to heat oil in it.

Unit 10 Kitchen Equipment and Tools 厨房设备和厨房工具

Jack: And then?

Chef: We'll roast chicken.

Jack: I see. And what shall I do?

Chef: Put the roasting tray in the oven.

Words & Expressions

tray [trei] *n.* 盘子，托盘

heat [hi:t] *v.* 使温暖，使热

roasting ['rəustiŋ] *adj.* 用于烤炙（烘焙）的

Drills

Drill One

A: What are you going to do with
- the rolling pin?
- the mincer?
- the peeler?
- the garlic press?

B: I'm going to
- flatten the dough.
- mince the beef.
- peel the potatoes.
- crush the garlic.

Drill Two

A: What is the
- boning knife
- oyster knife
- cleaver
- paring knife

for?

B: It is used to
- remove the meat from the bones.
- open oyster shells.
- chop the meat.
- peel fruits and vegetables.

 烹饪厨房英语

Exercises

Ⅰ. Fill in the blanks with correct letters.

kn_fe   st_ve   gl_ss
pee_er   p_n   pl_te
b_wl   chopst_cks   fo_k
table sp_ _n   t_ _th pick   n_pkin
chopping b_ _ _d   coffee p_t o_en

Ⅱ. Match the phrases and then make a sentence orally.

Example : We will A in/with a B .

A
1. fry fish
2. beat eggs
3. flatten the pastry
4. steam buns
5. boil eggs
6. peel the potatoes
7. cook fish
8. cut beef
9. wash dishes
10. stew chicken

B
a. peeler
b. wok
c. dishwasher
d. cleaver
e. whisk
f. frying pan
g. egg boiler
h. rolling pin
i. casserole
j. steamer

Ⅲ. Ask and answer the following questions with a partner.

1. What is the use of a boning knife?
2. Do you know how to open oyster shells?
3. How do you flatten the pastry?
4. Do you have a rolling pin?
5. Who likes apple pies?

Ⅳ. Complete the following sentences and then read them aloud.

1. I want to _____ an apple pie.
2. Is your apple pie _____?
3. Let's flatten it _____ a rolling pin.
4. _____ your pastry ready?
5. Here is a _____ pin.

Unit 10 Kitchen Equipment and Tools 厨房设备和厨房工具

V. Translate the following sentences into English.

1. 用打蛋器打蛋。
2. 厨师长需要用砂锅做鸡汤。
3. 给你擀面杖。
4. 我该用哪种刀呢?
5. 当心!别弄伤你的手。

 Part D: Reading

Cooking Knives

Cooking knives come in many shapes and sizes and are designed for specific tasks. Selecting the knife you need depends on your culinary experience, the style of cooking you prefer and how much money you wish to spend. Here is a list of the common types of knives and the basic purpose behind their design:

Boning Knife: It has a 4 ~ 5" blade and is a flexible knife for separating meat from the bone. This knife is more efficient for this purpose than a rigid knife.

Chef's Knife: Sporting a 4 ~ 12" blade, these versatile knives serve a variety of purposes but are particularly suited to chopping and dicing.

Cleaver: With a strong 6" blade the cleaver chops through bones and joints. You can even use the flat side of the cleaver to crush peppercorns and other spices.

Fish Fillet: It has a 7" blade. This thin, long blade has great flexibility and is perfectly suited for filleting fish. It is a superior tool.

Paring Knife: It has a 3 ~ 4" blade and is the most commonly used knife in the kitchen. Variations of this knife which is designed for peeling small round fruits and vegetables.

Carving Knife: An 8 ~ 10" blade that is used for carving paper-thin slices of meats, fruits and vegetables.

Steak Knife: With a 5" blade this sharp knife cuts through thick steaks and chops. It is designed to be an integral part of any place setting.

There you have it. These are the most common knives that we use as tools in our everyday life.

Words & Expressions

shape [ʃeip] *n.* 形状,模型
select [si'lekt] *vt.* 选择,挑选
purpose ['pɜ:pəs] *n.* 意志,目的
design [di'zain] *v.* 设计,绘制
culinary ['kʌlinəri] *adj.* 厨房的,烹饪的
blade [bleid] *n.* 刀片,剑

75

 **烹饪厨房英语**

efficient [i'fiʃnt] *adj.* 有效率的
crush [krʌʃ] *vt.* 压破，压碎
integral ['intigrəl] *adj.* 完整的
superior [su:'piəriə（r）] *adj.* （在质量等方面）较好的
depend on 依靠，依赖

chop [tʃɔp] *v.* 砍，伐，劈
peppercorn ['pepəkɔ:n] *n.* 胡椒粒
fillet ['filit] *n.* 肉片，鱼片

Unit 11

Chinese Pastry
中式面点

You will be able to:

1. Remember the English words of Chinese pastry.
2. Remember the useful verbs of processing flour.
3. Grasp the useful expressions about Chinese pastry in the dialogues.

烹饪厨房英语

Part A: Chinese Pastry

steamed stuffed bun
包子

boiled dumpling
水饺

steamed dumpling
蒸饺

wonton
馄饨

handmade noodle
手擀面

steamed bread
馒头

steamed roll
花卷

triangle bun stuffed with sugar 糖三角

fried dumplings
煎饺

fried leek dumplings
韭菜盒子

deep-fried pancake
油饼

baked scallion pancake 葱油饼

deep-fried glutinous rice balls with sesame 炸麻球

twist of dough
麻花

deep-fried dough stick 油条

Beijing-style cakes 京式糕点

Tangyuan (glutinous rice balls) 汤圆

small steamed bun 小笼包子

spring pancake
春饼

sweetheart cake
老婆饼

Words & Expressions

steamed [sti:md] *adj.* 蒸熟的
bun [bʌn] *n.* 小面包
triangle ['traiæŋgl] *n.* 三角（形）
baked [beikt] *adj.* 烤的，烘焙的
sesame ['sesəmi] *n.* 芝麻
dough [dəu] *n.* 生面团

stuffed [stʌft] *adj.* 塞满了的
wonton [ˌwɔn'tɔn] *n.* 馄饨
pancake ['pænkeik] *n.* 薄煎饼
glutinous ['glu:tənəs] *adj.* 黏性的
twist [twist] *n.* 拧，扭曲

Unit 11 Chinese Pastry 中式面点

Part B: English terms about flour

Ⅰ. Flour classification.

1. bread flour   高筋面粉
2. plain flour   中筋面粉
3. cake flour   低筋面粉

Ⅱ. English terms of processing flour.

1. make dough   和面
2. knead dough   揉面
3. twist the dough into a rope-like strip   面粉搓条
4. roll skin   擀皮
5. mix the flour   搅拌面粉
6. dough   面团

Words & Expressions

flour ['flauə(r)] n. 面粉，粉末　　　plain [plein] adj. 纯的，朴素的
dough [dəu] n. 生面团　　　knead [ni:d] v. 揉，捏制
twist [twist] v. 捻，搓，拧　　　rope-like adj. 绳索状的
strip [strip] n. 长条，条状　　　roll [roul] v. 滚动
mix [miks] v. 混合

Part C: Dialogues

Dialogue 1

Jack: Good morning, chef. What are you doing?

Chef: I'm making Jiaozi (dumpling).

Jack: May I help you?

Chef: Yes, of course. Jiaozi is not only delicious to eat but also easy to make.

Jack: Chef, could you tell me why Chinese like eating Jiaozi so much?

Chef: OK. Jiaozi is the traditional food in China. The appearance of Jiaozi looks like the gold or silver ingot used as money in ancient China.

烹饪厨房英语

　　As the Spring Festival marks the start of a new year, people choose to eat Jiaozi to wish for good fortune in the new year.

Jack: Oh, I see.

Words & Expressions

traditional [trə'dɪʃənl] *adj.* 传统的　　　appearance [ə'pɪərəns] *n.* 外表
ingot ['ɪŋɡət] *n.* 锭，铸块　　　　　　　ancient ['eɪnʃənt] *adj.* 古老的，古代的
mark [mɑ:rk] *v.* 标记　　　　　　　　　fortune ['fɔ:rtʃən] *n.* 命运，财产，运气

Notes:

1. Jiaozi is not only delicious to eat but also easy to make. 饺子不仅美味，而且容易制作。
   not only...but also... 不仅……而且……
2. The appearance of Jiaozi looks like the gold or silver ingot used as money in ancient China. 饺子的外形看起来就像古代中国用的金元宝或银元宝一样。
   look like... 看起来像……

Dialogue 2

A: What can I do for you, Madam!
B: I'd like to buy two boxes of Chinese moon cakes.
A: All right. We have the Guangdong style and the Suzhou style. Which one do you like?
B: What's the difference between them?
A: The skin of the Guangdong style cake is sweet, soft, puffy, palatable and filled with heavy stuffing. And The skin of Suzhou style cakes is puffy and filled with fragrant nuts.
B: The Suzhou style.
A: OK. Anything else?
B: Nothing more!

Words & Expressions

style [staɪl] *n.* 风格　　　　　　　　　skin [skɪn] *n.* 外皮
puffy ['pʌfi] *adj.* 膨胀的　　　　　　　palatable ['pælətəbl] *adj.* 美味的
stuffing ['stʌfɪŋ] *n.* （食品）填塞物　　fragrant ['freɪɡrənt] *adj.* 芳香的
nut [nʌt] *n.* 坚果，果仁

Notes:

1. What's the difference between them? 它们的区别是什么？
2. The skin of the Guangdong style cake is sweet, soft, puffy, palatable and filled with heavy

stuffing. 广式月饼的特点是外面的皮甜、软而松、味美、馅儿多。
3. And the skin of Suzhou style cakes is puffy and filled with fragrant nuts. 苏式月饼外面的皮松脆，馅儿味芳香。

Drills

Drill One

A: How do I...
- deep-fried pancake?
- steamed bread?
- sweetheart cake?
- dumpling?

B:
- First, ...
- Then, ...
- ...
- Finally, ...

Drill Two

A: Chef, could you tell me how to
- fry
- steam
- bake
- make
...?

B: OK.

Exercises

Ⅰ. Fill in the blanks with correct letters.

薄煎饼 p_ _cake   面团 do_gh
面粉 fl_u_   饺子 d_m_ling
馄饨 w_nt_n   面条 n_ _dle

Ⅱ. Match the correct words.

1. steamed bread   A. 莲蓉
2. sliced noodles   B. 包子
3. steamed stuffed bun   C. 韭菜盒子
4. bread flour   D. 老婆饼
5. lotus paste   E. 刀削面

烹饪厨房英语

6. glutinous rice   F. 馄饨
7. biscuit   G. 糯米
8. sweetheart cake   H. 高筋面粉
9. fried leek dumplings   I. 馒头
10. wonton   J. 饼干

Ⅲ. Fill in the blanks with the words given below. And then read the sentences aloud.

> make   knead   roll   mix   twist

1. How to _____ bread dough?
2. We should add water and _____ the flour.
3. You should _____ the dough into a rope-like strip.
4. We should _____ dough before making pastries.
5. _____ skin is not very difficult.

Ⅳ. Translate the following sentences into English.

1. 饺子不仅好吃，而且容易制作。
2. 饺子是中国的传统食物。
3. 我们有广式月饼和苏式月饼。您想要哪一种？
4. 苏式月饼皮松、馅香。
5. 制作面点前要先和面。

Ⅴ. Pair Work.

Make a dialogue about making Chinese pastry with your partner. Then role-play the dialogue, please.

Part D: Reading

Varieties of Rice Cakes

The variety of new year cakes varies according to customs of different regions. New year cakes have different flavor in the north and south. Northern cakes are steamed or fried, both are sweet. The South besides steamed and fried, there are pieces fried and soup cooked and so on with sweet or salty taste.

Unit 11 Chinese Pastry 中式面点

The North-style Rice Cake

Beijing new year cake is representative for the northern with yellow or white appearance symbolizing gold and silver which has the connotation of raising oneself higher in each coming year（年年高升）. Beijing cake includes jujube new year cake（枣年糕）, and white new year cakes made of glutinous rice or proso millet.

The South-style Rice Cakes

New year cakes in different regions of the South is also diversified. New year cakes cooking can be steamed, fried, sliced fried or soup. In Jiangsu, Suzhou-style rice cakes is a typical cake made of glutinous rice, mainly famous for osmanthus sugar cakes（桂花糖年糕） and lard cakes（猪油年糕）. In Zhejiang, Ningbo Cicheng's new year cakes mostly made of japonica rice is the most famous and the common practice is shredded pork & pickled cabbage rice cake（雪菜肉丝汤年糕）, Shepherd's-purse stir-fried with rice cake（荠菜炒年糕） and so on. In Shanghai, fried new year cake with pork rib（排骨年糕） is unique.

Words & Expressions

variety [vəˈraiəti] n.　多样，种类
region [ˈri:dʒən] n.　地区，范围
representative [ˌrepriˈzentətiv] adj.　典型的
connotation [ˌkɔnəˈteiʃn] n.　含义
proso millet [ˈprəusəu ˈmilit]　黍，黄米
osmanthus [ɔzˈmænθəs] n.　桂花
japonica [dʒəˈpɔnikə] rice　粳米
pickled [ˈpikld] adj.　腌制的，腌渍的
unique [juˈni:k] adj.　独特的，独一无二的

vary [ˈveəri] v.　改变，变化
flavor [ˈfleivə] n.　风味，风格
symbolize [ˈsimbəlaiz] v.　象征，用记号表现
glutinous [ˈglu:tənəs] rice　糯米
diversified [daiˈvə:sifaid] adj.　多样化的
lard [lɑ:d] n.　猪油
shred [ʃred] v.　撕成碎片
shepherd's-purse [ˈʃepədzpˈɜ:s] n.　荠菜

Unit 12

Western Pastry
西式面点

You will be able to:

1. Remember the English words of western pastry.
2. Remember the useful verbs of processing pastry.
3. Grasp the useful expressions in the dialogues.

Unit 12 Western Pastry 西式面点

🍲 Part A: Pastry Products

Bread 面包

pastry	toast	yeast bread	black bread
油酥面包	面包片，吐司	酵母面包	黑面包

hard roll	soft roll	muffin	bun
硬质面包	软质面包	松饼	小圆面包

Cakes 蛋糕

cake	cupcake	cream cake	cheese cake
蛋糕	小蛋糕	奶油蛋糕	奶酪蛋糕

pound cake	sponge cake	angel food cake	jelly cake
磅蛋糕	海绵蛋糕	天使蛋糕	果汁蛋糕

Pies 派

pie	banana cream pie	cherry pie	apple pie
派	香蕉奶油派	樱桃派	苹果派

lemon pie	chocolate cream pie	pumpkin pie	blueberry pie
柠檬派	巧克力奶油派	南瓜派	蓝莓派

烹饪厨房英语

Snacks 点心

tart 蛋挞　　　cookie 曲奇　　　biscuit 饼干

Cream 奶油点心

bavarian cream　vanilla ice cream　chocolate ice cream　strawberry ice cream
巴伐利亚奶油　　香草冰激凌　　　巧克力冰激凌　　　草莓冰激凌

vanilla pudding　milk pudding　fruit pudding　raisin pudding
香草布丁　　　　牛奶布丁　　　水果布丁　　　葡萄干布丁

Mousses 慕斯

chocolate mousse　green tea mousse　cream mousse　fruit mousse
巧克力慕斯　　　　绿茶慕斯　　　　奶油慕斯　　　水果慕斯

Words & Expressions

yeast [ji:st] n. 酵母　　　　　　　　muffin ['mʌfin] n. 松饼
bun [bʌn] n. 小圆面包　　　　　　　sponge [spʌndʒ] cake 海绵蛋糕
bavarian [bə'vɛəriən] adj. 巴伐利亚的

Part B: English terms of processing western pastry

knead the dough　　　　　　　　揉面团
fold the dough　　　　　　　　　折叠面团
shape the dough　　　　　　　　将面团塑形
flatten the dough　　　　　　　　擀平面团

Unit 12 Western Pastry 西式面点

mix the ingredients	混料
braid the dough	将面团做成麻花状
wrap the dough tightly	卷紧面皮
refrigerate it for at least thirty minutes	至少冷冻30分钟
divide the dough into four portions	把面团分成4个部分
roll the dough from the centre out	从中间向外擀面团

Words & Expressions

knead [ni:d] *vt.* 揉（面粉和水）成团
shape [ʃeip] *vt.* 做成某物的形状
ingredient [in'gridiənt] *n.* 成分，配料
wrap [ræp] *vt.* 包，裹，卷
refrigerate [ri'fridʒəreit] *vt.* 冷藏（食物）
portion ['pɔ:ʃn] *n.* 部分，一份

dough [dəu] *n.* 面团
flatten ['flætn] *v.* 使平，变平
braid [breid] *vt.* 把（头发）等编成辫子
tightly ['taitli] *adv.* 紧紧地
divide [di'vaid] *vt.* 分开，分割
roll [rəul] *vt.* 把……卷成筒状，滚动，翻滚

Part C: Dialogues

Dialogue 1

Waiter: Are you ready to order now?

Jack: Yes. I'd like to order something sweet. What do you have? Do you have any puddings?

Waiter: Of course. May I suggest chocolate pudding?

Jack: OK. I'll follow your suggestion.

Waiter: Thanks a lot. Wait a moment please.

Words & Expressions

sweet [swi:t] *adj.* 甜的
follow ['fɔləu] *vt.* 跟随

pudding ['pudiŋ] *n.* 布丁
suggestion [sə'dʒestʃən] *n.* 建议

Notes:

1. I'd like to order... 我打算点……
2. something sweet 一些甜点
3. follow one's suggestion(s) 听某人的建议

Dialogue 2

Waiter: Would you like to choose a birthday cake?

 烹饪厨房英语

Guest: Yes. It's my son's birthday. I would like to order a birthday cake.

Waiter: What kind of cake would you like? Would you prefer an almond cake or a chocolate cake?

Guest: A chocolate cake.

Waiter: What would you like written on the cake?

Guest: Let's see. Happy birthday, David! That's all.

Waiter: How many candles would you like on the cake?

Guest: Eight.

Waiter: How many servings of the chocolate cake do you need?

Guest: Six.

Waiter: When do you want the cake?

Guest: At noon tomorrow. Please send the cake up to Room 1608 at exactly 12:30 tomorrow. Thank you very much.

Waiter: My pleasure.

Words & Expressions

prefer [pri'fə:] *v.* 宁愿，更喜欢　　　　　　　exactly [ig'zæktli] *adv.* 确切地

serve [sə:v] *vt.* 为……服务，摆出（饭菜等）

Notes:

1. prefer 的用法有两种

（1）prefer to do sth. 更喜欢　例如：I prefer to stay at home on Sundays.

（2）prefer doing A to doing B 喜欢做A胜于做B

On rainy days, I prefer staying at home to going out.

prefer A to B 和B比起来更喜欢A

I prefer tea to coffee. 和咖啡比起来我更喜欢茶。

2. What would you like written on the cake? 请问您想要在蛋糕上写什么呢？

3. How many servings of the chocolate cake do you need?

请问巧克力蛋糕是几位要享用呢？

4. Please send the cake up to Room 1608 at exactly 12:30 tomorrow.

请你明天准时在12点30分把蛋糕送到1608房间。

Drills

Drill One

A: Is there anything I can do for you, Madam?

B: Yes, I'd like some bread for my friends.

Unit 12 Western Pastry 西式面点

A: How about some
- toast?
- yeast bread?
- black bread?
- hard roll?
- soft roll?

Drill Two

A: What tool shall we prepare?

B: We need
- a rolling pin.
- a cutting board.
- a pie plate.
- an egg beating whisk.

Exercises

Ⅰ. Match the correct words.

1. vanilla pudding   A. 酵母面包
2. fruit mousse   B. 蛋挞
3. muffin   C. 海绵蛋糕
4. yeast bread   D. 小圆面包
5. cheese cake   E. 油酥面包
6. lemon pie   F. 硬质面包
7. sponge cake   G. 松饼
8. hard roll   H. 香草布丁
9. pastry   I. 水果慕斯
10. green tea mousse   J. 磅蛋糕
11. tart   K. 柠檬派
12. soft roll   L. 绿茶慕斯
13. pound cake   M. 南瓜派
14. pumpkin pie   N. 奶酪蛋糕
15. bun   O. 软质面包

 烹饪厨房英语

Ⅱ. Complete the following conversation by choosing the appropriate expressions from the choices given. Then do the role-play with your partner. One will be the guest, the other will be the waiter or waitress.

W: How is the food, sir?

G: _____. Thank you.

W: _____ anything else?

G: Yes, I'd like to have some dessert. Have you got lemon pies?

W: I am sorry, sir. _____. _____ ? They are very tasty.

G: Well, then _____ two blueberry pies and one vanilla ice cream.

W: Yes, sir. Two blueberry pies and one vanilla ice cream.

Here are the choices:

A. make it

B. How about trying some blueberry pies

C. Would you like

D. I'm afraid we haven't got any more lemon pies

E. Very delicious

Ⅲ. Fill in the blanks with the important verbs we have learnt.

1. _____ the dough   折叠面团
2. _____ the dough   揉面团
3. _____ the dough   擀平面团
4. _____ the ingredients   混料
5. _____ the dough   将面团做成麻花状
6. _____ the dough tightly 卷紧面皮
7. _____ it for ten minutes 冷冻10分钟
8. _____ the dough into six portions 把面皮分成6个部分
9. _____ the dough from the centre out 从中间向外擀面团
10. _____ the dough   将面团塑形

Ⅳ. Translate the following sentences into English.

1. 我想吃些甜点，有蓝莓派吗？
2. 我建议您来份蛋挞或绿茶慕斯。
3. 就听你的。
4. 请问什么时候要蛋糕？
5. 我想为我的母亲订一个生日蛋糕。

Unit 12 Western Pastry 西式面点

🍲 Part D: Reading

The Origin of Custard Tart

The custard tart is regarded as one of Britain's traditional dishes. It remained a favourite over the centuries and is just as popular today. Even the name has an ancient pedigree — it is derived from both the old French word for crust, and the Anglo-Norman "crustarde", which meant a tart or pie with a crust.

Custard tarts were popular during Medieval times and records show that King Henry enjoyed them at his coronation banquet in 1399.

Today, custard tarts are usually made from short crust pastry, eggs, sugar, milk or cream, and vanilla, sprinkled with nutmeg and baked. However, in King Henry's day they could have included ingredients such as pork mince or beef marrow, but they were always filled with a sweet custard.

To make them a special treat they can be topped with fresh strawberries or raspberries. In the past they were much richer, and more fattening. Almond milk and honey was often added — much too sweet for today's tastes.

The custard tart is a lovely treat and goes wonderfully with a cup of Earl Grey tea. You just can't get more English than that.

Words & Expressions

origin ['ɔridʒin] n. 起源，由来
pedigree ['pedigri:] n. 家族，家谱
crust [krʌst] n. 面包皮，硬壳
Medieval [ˌmedi'i:vəl] times 中世纪时代
coronation [ˌkɔrə'neiʃn] n. 加冕礼
nutmeg ['nʌtmeg] n. 肉豆蔻
marrow ['mærəu] n. 髓，骨髓
raspberry ['rɑ:zbəri] n. 木莓，山莓

custard ['kʌstəd] n. 牛奶蛋糊（或冻）
derive [di'raiv] vt. 得来，得到
Anglo-Norman ['æŋgləu'nɔ:mən] adj. 诺曼系英国人的
sprinkle ['spriŋkl] vt. 撒（某物）于（某物之表面）
treat [tri:t] n. 款待，招待
a cup of Earl Grey tea 一杯伯爵茶

Unit 13

Chinese Cuisine
中　餐

You will be able to:

1. Remember the English words of seasonings.
2. Remember the eight kinds of Chinese cuisines and the traditional dishes.
3. Grasp the useful expressions about ordering dishes in the restaurants.

Unit 13 Chinese Cuisine 中餐

🍲 Part A: Seasoning

Words & Expressions

vinegar ['vinigə(r)] *n.* 醋   sesame oil ['sesəmi ɔil] *n.* 芝麻油
starch [stɑːtʃ] *n.* 淀粉   soybean paste ['sɔi,biːn peist] *n.* 大豆酱
scallion ['skæliən] *n.* 葱   star anise [stɑː 'ænis] *n.* 八角
caraway ['kærəwei] *n.* 香菜   cumin ['kʌmin] *n.* 孜然

🍲 Part B: Eight Kinds of Cuisine

Chinese Cuisine originated from the various regions of China and has become widespread in many other parts of the world. China has eight kinds of cuisine of special local flavors. Shandong, Sichuan, Jiangsu, Guangdong, Zhejiang, Fujian, Hunan and Anhui cuisines are all quite representative.

烹饪厨房英语

Shandong cuisine is pure and not greasy. Famous dish is Fried Pork Mixed with Sweet and Sour Sauce.

Sichuan cuisine is hot, spicy and sour. Famous dish is Mapo Tofu.

Jiangsu cuisine is salty and sweet, and quite delicious. Famous dish is Nanjing Pressed Duck.

Guangdong cuisine is sour, sweet, bitter, hot and salty flavoring. Famous dish is Roasted Suckling Pig.

Zhejiang cuisine is fresh, tender, soft and not greasy. Famous dish is Fried Shrimps with Longjing Tea.

Fujian cuisine tastes light and delicious and the color is beautiful. Famous dish is Fish Balls in Clear Soup.

Hunan cuisine is mostly seasoned with chili, and features sour, hot, delicious and fragrant tastes. Famous dish is Chicken with Chili and Pepper.

Anhui cuisine preserves most of the original taste and nutrition of the materials. Generally the food here is slightly spicy and salty. Famous dish is Stewed Soft-shelled Turtle with Ham.

Words & Expressions

cuisine [kwɪ'ziːn] *n.* 菜肴
representative [,reprɪ'zentətɪv] *adj.* 典型的
greasy ['griːsi] *adj.* 油腻的
spicy ['spaɪsi] *adj.* 辛辣的
salty ['sɔːlti] *adj.* 咸的
delicious [dɪ'lɪʃəs] *adj.* 美味的
tender ['tendə(r)] *adj.* 嫩的
taste [teɪst] *n. & v.* 味觉，品尝
fragrant ['freɪɡrənt] *adj.* 喷香的，香的

flavor ['fleɪvə] *n.* 味，特点
pure [pjʊə(r)] *adj.* 纯正的
hot [hɒt] *adj.* 辣的，热的
sour ['saʊə(r)] *adj.* 酸的
sweet [swiːt] *adj.* 甜的
bitter ['bɪtə(r)] *adj.* 苦的
soft [sɒft] *adj.* 温和的
chili ['tʃɪli] *n.* 红辣椒
nutrition [njuː'trɪʃn] *n.* 营养

Notes:

1. Fried Pork Mixed with Sweet and Sour Sauce 糖醋里脊

    sweet and sour sauce 糖醋

    例如：Chicken with Sweet and Sour Sauce 糖醋鸡块

2. Mapo Tofu 麻婆豆腐

3. Nanjing Pressed Duck 南京板鸭

Unit 13 Chinese Cuisine 中餐

4. Roasted Suckling Pig 烤乳猪
5. Fried Shrimps with Longjing Tea 龙井虾仁
6. Fish Balls in Clear Soup 清汤鱼丸
7. Chicken with Chili and Pepper 辣子鸡
8. Stewed Soft-shelled Turtle with Ham 火腿炖甲鱼

Part C: Dialogues

Dialogue 1

Waiter: Hello, ladies and gentleman! May I take your order now?
Guest: Yes, we want to try some dishes in Sichuan style. What would you recommend to us?
Waiter: What about Pork Lungs in Chili Sauce and Mapo Tofu?
Guest: Is it typical Sichuan flavor?
Waiter: Of course. Our chefs are from Sichuan.
Guest: OK. We will try them.

Words & Expressions

order ['ɔːdə(r)] n. 命令，秩序，点菜   recommend [ˌrekə'mend] v. 推荐
typical ['tipikl] adj. 典型的，代表性的

Notes:
1. May I take your order now? 您可以点菜了吗？
2. Pork Lungs in Chili Sauce 夫妻肺片

Dialogue 2

Mrs. Li: Have you ever been to Northeast China, Mrs. Liu?
Mrs. Liu: Of course. I'm from Harbin and I lived there for nearly 20 years.
Mrs. Li: Oh, great! You know, I am going there for a holiday this summer. Can you give me some ideas about the food there?
Mrs. Liu: Certainly. Northeast food is in a heavily seasoned taste. The main dishes are Hot Candied Sweet Potato, Mutton Shashlik, Stewed Chicken with Mushroom, Braised Pork with Vermicelli and so on.
Mrs. Li: How about the price?

Mrs. Liu: Not very expensive, but the service is quite good. And snack bars can be seen everywhere.

Mrs. Li: Really? I can't wait to go there!

Words & Expressions

price [prais] *n.* 价格，价钱　　expensive [ik'spensiv] *adj.* 昂贵的

service ['sə:vis] *n.* 服务　　snack bar 小吃店

Notes:

1. Northeast food is in a heavily seasoned taste. 东北菜口味重。
2. Hot Candied Sweet Potato 拔丝地瓜
3. Mutton Shashlik 羊肉串
4. Stewed Chicken with Mushroom 小鸡炖蘑菇
5. Braised Pork with Vermicelli 猪肉炖粉条
6. I can't wait to go there! 我等不及要去那儿了！

Drills

Drill One

A: May I take your order now?

B: Yes, we want to try
- some dishes in Sichuan style.
- some dishes in Guangdong style.
- some dishes in Shandong style.
- some dishes in Zhejiang style.

Drill Two

A: What would you recommend to us?

B: What about
- Pork Lungs in Chili Sauce?
- Roast Suckling Pig?
- Sauted Potato, Green Pepper and Eggplant?
- Begger's Chicken?

Exercises

Ⅰ. Match the correct words.

1. salt　　　　A. 味精
2. pepper　　　B. 糖

Unit 13 Chinese Cuisine 中餐

3. vinegar   C. 酱油
4. sugar   D. 姜
5. soy sauce   E. 淀粉
6. garlic   F. 盐
7. starch   G. 蒜
8. oil   H. 醋
9. MSG   I. 胡椒粉
10. ginger   J. 油

II. Match the correct words.

1. The food is light, fresh and sweet.   A. Sichuan food
2. The food is fresh, tender and fragrant.   B. Guangdong food
3. The food is salty, fresh and fragrant.   C. Hunan food
4. The food is sweet and sour.   D. Shandong food
5. The food is very spicy.   E. Jiangsu food
6. The food is light and clear.   F. Zhejiang food
7. The food is sour and hot.   G. Anhui food
8. The food is fresh, tender and spicy.   H. Fujian food

III. Complete the following dialogue and read it aloud.

Waiter: Here you are, sir. Here is the menu.
Guest: Oh, thanks. What's good tonight?
Waiter: We have very good dishes for tonight, which do you prefer, Guangdong food or Sichuan food?
Guest: _____（我想尝尝广东菜）
        _____（你推荐些什么？）
Waiter: _____（怎么样）Fried Crisp Pork and Roasted Suckling Pig?
Guest: _____（听起来不错）Give me the dishes, please.

IV. Translate the following sentences.

1. 中国有八大菜系。
2. 四川菜又酸又辣又麻。
3. 先生，您现在可以点菜了吗？
4. 东北菜口味比较重。
5. 这家餐厅的服务非常好。
6. Jiangsu cuisine is salty and sweet, and quite delicious.

7. Is it typical Sichuan flavor?

8. Can you give me some ideas about the food there?

V. Which dishes are real northeastern food?

1. Fried Pork in Scoops

2. West Lake Fish in Vinegar Gravy

3. Hot Candied Sweet Potato

4. Fried Carp with Sweet and Sour Sauce

5. Nanjing Pressed Duck

6. White Meat with Pickled Chinese Cabbage

7. Braised Pork with Vermicelli

8. Chongqing Hot Pot

Part D: Reading

Why Sichuan Food Is So Hot?

It is known to all that Sichuan food is peppery, but tasty. Do you know why Sichuan food is so hot?

Sichuan is in the west of China. The weather there is damp and cold because of high altitude. People in such weather often suffer from rheumatism. It can cause heart disease, pains in joints and so on. In order to prevent the diseases, the Sichuanese always take lots of spicy food to perspire. That's why people in Sichuan prefer hot and spicy food.

Why do so many people in other parts of the country enjoy Sichuan food? The answer is quite simple: The dishes are very delicious. Now Sichuan cuisine has become one of the eight traditional Chinese cuisines.

Words & Expressions

peppery ['pepəri] adj. 胡椒味的
tasty ['teisti] adj. 美味的，可口的
damp [dæmp] adj. 潮湿的
altitude ['æltitju:d] n. 海拔
suffer ['sʌfə(r)] v. 患病
disease [di'zi:z] n. 疾病
joints [dʒɔint] n. 关节
perspire [pə'spaiə(r)] v. 出汗，流汗
traditional [trə'diʃənl] adj. 传统的

Unit 14

Western Cuisine
西 餐

You will be able to:

1. Remember the English words of western seasonings.
2. Grasp the useful expressions about ordering lunch and buffet.
3. Know the table manners in western countries.

烹饪厨房英语

Part A: Seasoning

Ⅰ. 西餐常用调味品及调味用酒

| worcestershire 辣酱油 | tomato paste 番茄酱 | tomato ketchup 番茄沙司 | curry 咖喱 |

| mustard 芥末 | wine vinegar 葡萄酒醋 | apple cider vinegar 苹果醋 | white vinegar 白醋 |

| salt 盐 | pepper powder 胡椒粉 | chicken bouillon 鸡粉 | beef stock 牛肉粉 |

| brandy 白兰地 | red wine 红葡萄酒 | white wine 白葡萄酒 | port wine 波尔图酒 |

Ⅱ. 西餐常用香、辛调料

| bay leaf 香叶 | pepper 胡椒 | parsley 番芫荽 | clove 丁香 |

| mint 薄荷叶 | thyme 百里香 | rosemary 迷迭香 | sage 鼠尾草 |

Unit 14 Western Cuisine 西餐

   basil 罗勒   dill 莳萝   saffron 藏红花   marjoram 牛膝草

   oregano   tarragon   citronella   paprika
   小茴香   他拉根香草   香茅   红椒粉

Ⅲ. 调味汁及沙司

brown sauce 布朗沙司	cream sauce 奶油沙司
red wine sauce 红酒沙司	rosemary sauce 迷迭香沙司
orange sauce 鲜橙沙司	pepper sauce 胡椒沙司
Hollandaise sauce 荷兰沙司	French dressing 法国汁
curry sauce 咖喱沙司	Paris butter 巴黎黄油
snail butter 蜗牛黄油	mayonnaise sauce 马乃司沙司
tartar sauce 鞑靼沙司	caviar sauce 鱼子酱沙司
tomato sauce 番茄沙司	thousand islands dressing 千岛汁

Words & Expressions

worcestershire ['wustəʃiə] n. 辣酱油
mustard ['mʌstəd] n. 芥末
parsley ['pɑ:sli] n. [植]西芹，欧芹，番芫荽
saffron ['sæfrən] n. [植]藏红花
citronella [ˌsitrə'nelə] n. 香茅
caviar ['kævia:（r）] n. 鱼子酱

ketchup ['ketʃəp] n. 番茄酱
bouillon ['bu:jɔn] n. 肉汤，牛肉汤
thyme [taim] n. （用以调味的）百里香（草）
marjoram ['mɑ:dʒərəm] n. 牛膝草
paprika [pæ'prikə] n. 红辣椒，辣椒粉

Part B: Dialogues

Dialogue 1

Chef: Jack, prepare some cold sauce for me.
Jack: OK. What cold sauce should I prepare?

 烹饪厨房英语

Chef: French sauce, Italian sauce, Thousand-island sauce, Mustard sauce and Cheese sauce.

Jack: I see.

Chef: Tom, prepare a fruit salad.

Tom: OK. What sauce shall I use? And what fruit? Is the pineapple OK?

Chef: The cream as the sauce. You can use the pineapple with some cherries and raisins.

Tom: OK. I'll get it ready soon.

Chef: Peter, what are you doing?

Peter: I'm making veal pie. Would you like to have a taste?

Chef: Very good except it's lightly seasoned. Get more salt, please. And the color should be deeper.

Peter: I see. Thanks for your advice.

Words & Expressions

raisin ['reizn] *n.*   葡萄干   mustard ['mʌstəd] *n.*   芥末，芥菜
cheese [tʃi:z] *n.*   奶酪   veal [vi:l] *n.*   小牛肉
taste [teist] *v. & n.*   尝；品尝   season ['si:zn] *n. & vt.*   季节；调味
French sauce   法国汁   Italian sauce   意大利汁
mustard sauce   芥末汁   cheese sauce   奶酪酱

Notes:

1. prepare some cold sauce 准备一些冷菜调味汁
   cold sauce 冷菜调味汁

2. Would you like to have a taste? 请您尝一尝好吗？

3. Very good except it's lightly seasoned. 非常好，就是有点淡。
   light指味道淡的，salty指咸的

Dialogue 2

W: Waitress 服务员（女）   G: Guest （客人）

W: Good evening. Welcome to our restaurant.

G: Good evening. I'm Mary Smith. I have a reservation.

W: This way, please. Your table is near the window.

G: Thank you.

Unit 14 Western Cuisine 西餐

W: Here is the menu. Take your time.
G: Thank you.
W: May I take your order now?
G: Yes. I'll have a fruit salad and this steak.
W: How would you like your steak cooked? Rare, medium or well-done?
G: Medium.
W: What would you like to go with your steak?
G: French fries, peas and carrots.
W: What would you like for dessert?
G: Ice cream, please. Could I have the check, please?
W: Here is the check.
G: Can I pay for the bill by credit card?
W: Yes, of course. Here's your receipt.

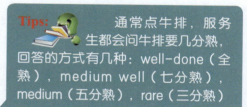

Tips: 通常点牛排，服务生都会问牛排要几分熟，回答的方式有几种：well-done（全熟），medium well（七分熟），medium（五分熟），rare（三分熟）

Words & Expressions

steak [steik] n. 牛排
French fries 薯条
pay [pei] v. 支付
dessert [di'zə:t] n. 甜点
credit card 信用卡

menu ['menju:] n. 菜单
reservation [ˌrezə'veiʃn] n. 预订
receipt [ri'si:t] n. 收据
bill [bil] n. 账单

Notes:

1. 在西餐厅就餐，菜单一般由3类不同的菜别组成。the first course 或者 starter 指第一道菜，往往是汤、一小碟沙拉或者其他小吃。第二道菜是 main course（主菜），常为肉食、鸡、鱼、牛羊肉、海鲜等，并配有蔬菜、蘑菇、土豆等。最后一道菜是 dessert（甜食），或冷或热，但都是甜的。西方饮食习惯是先吃咸的，后吃甜的。

2. I have a reservation. 我有预订。高级餐厅用餐前一般都需要预订。

3. Take your time. 请慢慢看。

4. May I take your order now? 您现在点菜吗？

5. What would you like to go with your steak? 您的牛排要配什么菜呢？
   go with... 与……相配
   例如：This hat doesn't go with my new dress. 这顶帽子和我的新裙子不搭配。

6. Could I have the check, please? 我想结账。

7. Can I pay for the bill by credit card? 我可以用信用卡付账吗？

 烹饪厨房英语

pay for the bill 付账

Dialogue 3

W: Waiter 服务员（男） G: Guest（客人）

G: Look at the dishes, they must taste very good.
W: Yes, madam. Today our restaurant serves buffet for you.
G: I do like buffet.
W: Come this way, please. Our buffet price today is 60 yuan for each one.
G: That's OK. What do you have here today?
W: Look at the buffet table, please. In the middle is the roast beef. It's excellent. Beside it there are Noodles with Eggs and Bacon, Fried Bacon, ham, sausage and so on.

G: Sounds good.
W: And by the window you can see some cold dishes on the table. They're salad, pickles and some vegetables. They're so fresh.
G: I can't wait to have a taste. Thank you so much for telling me this.
W: Enjoy your meal.

Words & Expressions

buffet ['bufei] n. 自助餐
bacon ['beikən] n. 培根
sausage ['sɔsidʒ] n. 香肠，腊肠
roast beef 烤牛肉

price [prais] n. 价格，价钱
ham [hæm] n. 火腿
pickle ['pikl] n. 腌菜，泡菜

Notes:

1. 自助餐在西方称为buffet，正规的解释为冷餐会，虽然吃自助餐比吃正规西餐自由一些，但也有它的规矩。上菜顺序一般是冷菜、汤、热菜、甜点、水果、冰激凌。

2. I do like buffet. 我非常喜欢吃自助餐。
   do和行为动词连用，起加强语气的作用。

3. I can't wait to have a taste. 我都等不及要尝一尝了。

4. Noodles with Eggs and Bacon 培根鸡蛋面

Unit 14 Western Cuisine 西餐

Drills

Drill One

A: What do you have for breakfast?

B: I'd like to have
- some eggs.
- some porridge.
- a cup of coffee.
- soybean milk（豆浆）.
- some noodles.
- pineapple juice.
- steamed buns.

Drill Two

A: May I take your order now?

B: Yes, I'll have
- steak.
- a fruit salad.
- fried fish.
- French fries.
- tomato soup.
- roast beef.
- Noodles with Eggs and Bacon.

Drill Three

A: How would you like your steak cooked?

B:
- Rare,
- Medium,
- Well-done,

please.

A: What would you like to go with your steak?

B: French fries, peas and carrots.

烹饪厨房英语

Exercises

Ⅰ. Match the correct words.

1. mustard   A. 莳萝
2. curry   B. 芥末
3. rosemary   C. 罗勒
4. dill   D. 百里香
5. bay leaf   E. 咖喱
6. basil   F. 迷迭香
7. French dressing G. 布朗沙司
8. brown sauce   H. 丁香
9. thyme   I. 香叶
10. clove   J. 法国汁

Ⅱ. Complete the following dialogue and read it aloud.

> credit card   welcome   menu   well-done   reservation   go with   like   order

W: Good evening. _____1_____ to our restaurant.

G: Good evening. I'm Mary Smith. I have a _____2_____ .

W: This way, please. Your table is near the window.

G: Thank you.

W: Here is the _____3_____. Take your time.

G: Thank you.

W: May I take your _____4_____ now?

G: Yes. I'll have a fruit salad and this steak.

W: How would you like your steak cooked? Rare, medium or _____5_____ ?

G: Medium.

W: What would you like to _____6_____ your steak?

G: French fries, peas and carrots.

W: What would you _____7_____ for dessert?

G: Ice cream, please. Could I have the check, please?

W: Here is the check.

G: Can I pay for the bill by _____8_____ ?

W: Yes, of course. Here's your receipt.

Ⅲ. Translate the following sentences into English.

1. 您想要几分熟的牛排？五分熟？七分熟？还是全熟？

2. 我能用信用卡付账吗？
3. 我要一杯菠萝汁，两个煮得老一点的鸡蛋和两个吐司面包。
4. 您的牛排要配什么菜呢？
5. 这是菜单，请慢慢看。
6. 我等不及要尝一尝了。

IV. Role-play.

Mr. Bell and Mrs. Bell are having dinner in the hotel. A waiter is serving them. Mr. Bell likes fried fish and chips. Mrs. Bell likes fried beef steak with black pepper sauce. Make a dialogue between the waiter and the couple.

V. Discuss.

1. What have you learned about the serving of western cooking?
2. How much do you know about the manner of western cooking?

Part C: Reading

Table Manners in Western Countries

It's said that one can know the nature of both man and woman by their actions at table. Indeed table manners are closely related to one's life, one's job and one of signals of one's manners.

As the saying goes: "When in Rome, do as the Romans do." English learners should know what are the proper table manners in western countries if invited to a dinner party.

The followings are some suggestions that may be helpful to you.

The first important thing that you need to know is when to begin to eat. Before dinner, the hostess usually serves guests first and herself last. So do not eat as soon as you are served but wait until the hostess also has been served and has picked up her fork as the signal to begin. But at a large dinner party where there are many guests, the hostess may ask everyone to begin eating as soon as they are served.

If you want to talk while eating, your mouth should certainly not be full of food because it is considered very bad manners. But it is possible to speak with a little food in the mouth. When you have to answer a question, naturally you must wait until much of the food in your mouth is eaten.

When a dish is passed to you with a fork in the plate, you are supposed to use the fork to take your food. Do not take too much at a time to make sure that other guests can have enough food. So first please look around the table and count the number of guests, and then take your food accordingly.

Anyway, good table manners are very important and it is more complicated than we thought. But it is necessary for one to know what are good table manners because they can help one succeed in his life.

Words & Expressions

nature ['neɪtʃə(r)] *n.* 自然，天性
consider [kən'sɪdə(r)] *vt.* 考虑，认为
suggestion [sə'dʒestʃən] *n.* 建议，意见
signal ['sɪgnəl] *n.* 信号，暗号
succeed [sək'siːd] *v.* 成功
be related to... 与……有关，和……有联系
be supposed to... 应该
When in Rome, do as the Romans do. 入乡随俗。

serve [sɜːv] *vt.* 招待
fork [fɔːk] *n.* 餐叉
hostess ['həʊstəs] *n.* 女主人
complicated ['kɒmplɪkeɪtɪd] *adj.* 复杂的

Unit 15

Show Kitchen
明 档

You will be able to:
1. Know the characteristics of show kitchen.
2. Remember the English words of different show kitchens and the dishes in the hotel.
3. Grasp the useful expressions in the dialogues.

烹饪厨房英语

Part A: Introduction

Show kitchen: usually means a plate of raw materials display at the eye-catching place to attract customers. Many restaurants have open cooking areas. Guests and cook can talk directly, and then the cook cooks what the guests order.

The characteristics of show kitchen:

1. neat
2. bright
3. fresh materials
4. fine and even cutting skills
5. reasonable balance in diet
6. sanitation to every detail

Words & Expressions

raw [rɔ:] *adj.* 生的,未加工的
attract [ə'trækt] *vt.* 吸引
even ['i:vən] *adj.* 均匀的
sanitation [ˌsæni'teiʃn] *n.* 卫生系统或设备
detail ['di:teil] *n.* 详述,各种细节

display [di'splei] *n.* 展览,陈列,陈列品
area ['eəriə] *n.* 地区,区域,范围
neat [ni:t] *adj.* 整洁的,干净的
balance ['bæləns] *n.* 均衡

Part B: Different show kitchens and dishes

Western Kitchen（西式厨房）：

1. hamburger 汉堡包
2. spaghetti 意大利面
3. sandwich 三明治
4. penne 管面

Unit 15 Show Kitchen 明档

Western Grill（西式烧烤）：

1. Australian beef tenderloin
   澳大利亚牛柳
2. New Zealand lamb chops
   新西兰羊排
3. T-bone steak
   T骨牛排
4. Australian baby lobster
   澳大利亚龙虾仔
5. salmon fillet
   三文鱼柳

Asian Grill（亚式烧烤）：

1. chicken wings
   鸡翅
2. grilled king prawn
   扒大虾
3. lamb shashlik
   羊肉串

Teppanyaki（铁板烧）：

1. scallop
   扇贝
2. chicken drumsticks
   鸡腿
3. tiger prawn
   大虾
4. salmon
   三文鱼

111

Noodle Bar（面条吧）：

1. dan dan noodles
   担担面
2. wonton noodles
   云吞汤面
3. hand pulled noodles
   拉面
4. Beijing style bean paste noodles
   老北京炸酱面

Steam basket（蒸笼）：

1. steamed spring chicken
   童子鸡
2. steamed shrimp dumpling
   翡翠虾饺
3. steamed minced pork dumpling
   小笼包
4. steamed crab with glutinous rice
   糯米蒸蟹

Soup（汤）：

1. black chicken soup
   乌鸡汤
2. pumpkin soup
   南瓜汤
3. cream of mushroom soup
   奶油蘑菇汤

Unit 15 Show Kitchen 明档

Others（其他）：

1. sashimi
   刺身
2. sushi
   寿司
3. assorted fruit and vegetable salad
   什锦水果蔬菜沙拉

Words & Expressions

spaghetti [spə'geti] n. 意大利面条
tenderloin ['tendəlɔin] n. （牛、羊等的）腰部嫩肉，里脊
glutinous ['glu:tənəs] rice n. 糯米，黏米
drumstick ['drʌmstik] n. 鼓槌，鼓槌（形物）体（如鸡腿等）

penne ['penei] n. 管面
shashlik [ʃɑ:ʃ'lik] n. 烤肉串
bean paste [bi:n peist] 豆瓣酱
assorted [ə'sɔ:tid] adj. 各式各样的
sashimi [sɑ:'ʃi:mi] n. [日]生鱼片
sushi ['su:ʃi] n. [日]寿司

 Part C: Dialogues

Dialogue 1

Mary: Look! There are grilled meat, prawn and vegetable dishes over there.

Tom: Right! The chef cooks the food in front of the guests.

Mary: Great, let's go and have a look!

...

Cook: Good evening. What would you like to order?

Mary: I'd like the prawns, and I'd like the chicken wings, too.

Cook: OK. Then do you like hot and spicy food?

Mary: Yes, I do. What would you recommend?

Cook: This is a delicious prawn dish, with chilies and garlic.

Mary: That sounds good. I'll have that.

 烹饪厨房英语

Words & Expressions

spicy ['spaisi] *adj.* 辛辣的   recommend [ˌrekə'mend] *vt.* 推荐，劝告

prawn [prɔːn] *n.* 对虾，明虾   chili ['tʃili] *n.* 红辣椒

Notes:

1. I'd like the prawns. 我想要虾。

  I'd like... 我想要……

2. What would you recommend? 你能为我推荐一下吗？

Dialogue 2

Mum: Honey, what do you want for breakfast?

Child: I'd like eggs, Mummy.

Mum: OK.

...

Cook: Good morning. May I help you?

Mum: My child wants to eat eggs.

Cook: How would you like your eggs? Boiled, fried or scrambled?

Child: I'd like fried eggs.

Cook: How would you like me to cook your eggs? Sunny-side up or over easy?

Mum: Over easy.

Cook: OK. Just wait a moment, and please keep the child away from the teppan.

Mum: Thank you!

Words & Expressions

boil [bɔil] *vt.* 用开水煮   fry [frai] *vt.* 油炸，油煎

scramble ['skræmbl] *vt.* 炒（蛋）   sunny-side up 单面煎

over easy 双面煎   teppan 铁板

Notes:

1. How would you like your eggs? Boiled, fried or scrambled? 您想要什么样的鸡蛋？煮鸡蛋、煎鸡蛋还是炒鸡蛋？

2. Please keep the child away from the teppan. 请让您的孩子远离铁板。

  keep... away from... 使……远离……

  例如：Keep the child away from the fire. 让孩子离火远点。

Unit 15 Show Kitchen 明档

Drills

Drill One

A: What would you like to order?

B: I'd like
- hamburger.
- Beijing style bean paste noodles.
- T-bone steak.
- black chicken soup.
- sashimi.
- chicken wings.

Drill Two

A: How would you like your eggs? Boiled, fried or scrambled?

B: I'd like
- boiled
- fried
- scrambled

eggs.

Exercises

Ⅰ. Match the correct words.

1. sushi   A. 意大利面
2. sashimi   B. 三明治
3. sandwich   C. 云吞面
4. spaghetti   D. 沙拉
5. hamburger   E. 寿司
6. salmon fillet F. 南瓜汤
7. wonton noodles G. T骨牛排
8. T-bone steak   H. 汉堡包
9. salad   I. 刺身
10. pumpkin soup J. 三文鱼柳

Ⅱ. Complete the following dialogue and read it aloud.

smells   breakfast   scrambled   sunny   crazy   soft   picky

A: Wow, you're up early today! What's for _____?

115

 烹饪厨房英语

B: Well, I felt like baking, so I made some muffins.

A: _____ good! I'll make some coffee. Do you want me to make you some eggs?

B: Sure, I'll take mine _____ side up.

A: I don't know how you can eat your eggs like that!

B: You know, my dad had _____ eggs every morning for twenty years. It drove my mom _____!

A: You know what really drives me crazy? When I ask for _____ boiled eggs, and they overcook them, so they come out hard boiled!

B: You're so _____ sometimes.

Ⅲ. Translate the following sentences into English.

1. 很多酒店都有明档区。
2. 您想点什么？
3. 我想要烤鸡翅。
4. 晚上好，先生。我能帮助您吗？
5. 您想要什么样的鸡蛋？煮蛋还是煎蛋？
6. 让孩子离铁板远点。

Ⅳ. According to the dialogues, make up a new dialogue between the cook and the guest.

Ⅴ. Read the sentences loudly and try to recite.

1. Good morning/afternoon/evening, sir/madam.
2. May I help you?/ Can I help you?/ What can I do for you?
3. Welcome to our hotel/restaurant, sir/madam.
4. Nice to meet you, sir.
5. We're glad to have you here.
6. What would you like to order?
7. I'm sorry to have kept you waiting.
8. It doesn't matter./ Never mind.
9. Thank you very much.
10. You're welcome. /It's my pleasure.
11. Have a good time!
12. Have a nice trip!
13. Wish you a pleasant journey.
14. Hope to see you again.

Unit 15 Show Kitchen 明档

15. Enjoy your dinner.
16. Here is the menu.

Part D: Reading

Teppanyaki

Teppanyaki is a style of Japanese cuisine that uses an iron griddle to cook food. The word *teppanyaki* is derived from *teppan*, which means iron plate, and *yaki*, which means grilled, broiled, or pan-fried. In Japan, teppanyaki refers to dishes cooked using an iron plate, including steak, shrimp, *okonomiyaki*, and *yakisoba*.

Modern *teppanyaki* grills are typically propane-heated flat surface grills and are widely used to cook food in front of guests at restaurants. Teppanyaki grills are commonly confused with the *hibachi* barbecue grill, which has a charcoal or gas flame and is made with an open grate design. With a solid griddle type cook surface, the *teppanyaki* is more suitable for smaller ingredients, such as rice, egg, and finely chopped vegetables.

Words & Expressions

teppanyaki　铁板烧
yakisoba　日式炒面
derive [di'raiv] v.　从……中提取
flat [flæt] adj.　平的
barbecue ['bɑ:bikju:] n.　烤肉，烧烤野餐
flame [fleim] n.　火焰

okonomiyaki　日本烧饼
griddle ['gridl] n.　煎饼用浅锅
broil [brɔil] vt.　烤（焙、炙等）
hibachi [hi'bɑ:tʃi] n.　（日本）木炭火盆
suitable ['su:təbl] adj.　合适的，适当的
ingredient [in'gri:diənt] n.　（烹调的）原料

Unit 16

Hotel
酒 店

You will be able to:
1. Remember the English words of facilities of the hotel.
2. Remember the different positions in the hotel.
3. Fill in personal information form correctly.
4. Make a fluent self-introduction in the interview.

Part A: Hotel Facilities

1. Front Desk 前台
2. Business Center 商务中心
3. Coffee Shop 咖啡厅
4. Lobby 大堂
5. Lobby Bar 大堂吧
6. Music Bar 音乐吧
7. Nightclub 夜总会
8. Chinese Restaurant 中餐厅
9. Western Restaurant 西餐厅
10. Fitness Center 健身中心
11. Swimming Pool 游泳池
12. Snooker Room 桌球室

13. Card Room 牌艺室
14. Beauty Salon 美容美发厅
15. Gift Shop 礼品店
16. Cashier's 收银处
17. Function Room 多功能厅
18. Conference Room 会议室

烹饪厨房英语

Part B: Hotel Position

Words & Expressions

executive chef [ig'zekjətiv ʃef] n. 行政主厨　　　　pantryman ['pæntrimən] n. 司膳总管
relief [ri'li:f] cook n. 替班厨师　　　　　　　butcher ['butʃə（r）] n. 屠夫

Part C: Dialogues

Dialogue 1

Chef: Jack, I would like to introduce you to Mike. He is our Sauce Cook. Mike, this is

Jack, a commis.

Jack: Nice to meet you, Mike.

Mike: Nice to meet you too, Jack. Welcome to join us!

Jack: I am a new-comer. I hope you can help me.

Mike: No problem. Everyone starts from new. I hope you can enjoy your job.

Jack: I am sure I will.

Words & Expressions

chef [ʃef] *n.* 厨师长

new-comer [nju:'kʌmə（r）] *n.* 初学者，新手

introduce [ˌintrə'dju:s] *v.* 介绍

Dialogue 2

Tom: Where do you work?

Jack: I work in Shangrila Hotel.

Tom: What do you do?

Jack: I'm a cook.

Tom: Really? Then what do you do in the kitchen?

Jack: I'm a breakfast cook in charge of making breakfast.

Tom: Oh, sounds good. How long do you work every day?

Jack: 9 hours.

Tom: What do you think of your job?

Jack: I think it's very satisfying, when the guests enjoy the food I made, I feel very happy.

Tom: Great!

Words & Expressions

kitchen ['kitʃin] *n.* 厨房

in charge of 主管，负责

satisfying ['sætisfaiiŋ] *adj.* 令人满意的

Shangrila Hotel 香格里拉酒店

Notes:

1. What do you do? 你是做什么工作的？

2. How long do you work every day? 你每天工作多长时间？
   how long 多久

3. What do you think of your job? 你认为你的工作怎么样？
   I think it's very... 我认为我的工作很……
   例如：I think it's very boring/challenging/stressful.
   我认为我的工作非常令人厌烦/具有挑战性/压力很大。

Dialogue 3

Interviewer: Sit down, please.

Jack: OK.

Interviewer: Could you introduce yourself briefly?

Jack: Yes. I'm Jack. I'm 18 years old. My major is Chinese Cooking.

Interviewer: I see. Tell me the courses you have learned.

Jack: Chinese, Math, English, Chinese Cooking, Chinese Pastry, Carving, Culinary Art, Nutrition and so on.

Interviewer: What's your favorite subject?

Jack: I like Chinese Cooking and Carving best.

Interviewer: Why do you want to work for our hotel?

Jack: I know your hotel is a five-star hotel with a good reputation both at home and abroad. I'm desirous of working in a large hotel just like yours.

Interviewer: OK. Do you think you're qualified for this job?

Jack: I think I am. I'm hardworking, outgoing and cooperative. I can work overtime and get along well with others.

Interviewer: Good. Do you have any part-time work experience?

Jack: Yes, I once worked as a cook cutting vegetables in a restaurant during summer vacations.

Interviewer: I see. Do you have any questions?

Jack: When can I know the result?

Interviewer: In about one week. Just wait for our reply.

Words & Expressions

course [kɔː(r)s] *n.* 科目，进程
desirous [di'zaiərəs] *adj.* 渴望……的
experience [ik'spiəriəns] *n.* 经验
outgoing ['autgəuiŋ] *adj.* 开朗的
get along well with sb. 和某人相处融洽
reputation [ˌrepju'teiʃn] *n.* 名气
overtime ['əuvətaim] *adv.* 加班地
reply [ri'plai] *n.* 答复
cooperative [kəu'ɔpərətiv] *adj.* 合作的
summer vacation 暑假

Notes:

1. I know your hotel is a five-star hotel with a good reputation both at home and abroad.
   我知道贵酒店是一家五星级酒店，而且在国内外享有盛名。
   at home and abroad 国内外

2. Do you think you're qualified for this job? 你认为你能胜任这份工作吗？

be qualified for... 胜任（某种工作）

3. Do you have any part-time work experience? 你有没有兼职的工作经验？
part-time 兼职的

Drills

Drill one

A: Can you introduce me to Mike?

B: OK! Mike,
- this is Jack, a commis.
- this is night cook, Tom.
- this is fish cook, Amy.
- this is pastry chef, Tom.

Drill two

A: What do you do in the kitchen?

B: I'm a
- breakfast cook
- soup cook
- pantryman
- kitchen steward

in charge of
- making breakfast.
- making soup.
- meals.
- kitchens.

Drill three

A: Could you introduce yourself briefly?

B: Yes. I'm Jack. I'm 18 years old.

My major is
- Chinese Cooking.
- Chinese Pastry.
- Western Cooking.
- Western Pastry.

Exercises

Ⅰ. Choose the correct answer.

1. Mr. Wood is swimming in the _____.
   A. swimming pool   B. conference room

2. Mr. Brown will have dinner at the _____.
   A. business center   B. restaurant
3. Guests can buy something in the _____.
   A. gift shop   B. business center
4. Guests can have a cup of tea in the _____.
   A. lobby   B. coffee shop
5. Guests check in at the _____.
   A. front office   B. housekeeping
6. Guests' clothes will be washed in the _____.
   A. room service   B. laundry department

Ⅱ. Write down the names according to their duties.

1. _____ The person who can relieve everyone.
2. _____ The person who makes the breakfast.
3. _____ The person who puts the meat into the oven.
4. _____ The person who makes the dishes for the staff.
5. _____ The person who makes the soup.

Ⅲ. Fill in the Personal Information Form.

Full Name		Gender	
Address			
ID Card		Place of Birth	
Mobile		Personal E-mail	
Martial Status	Nationality		Postal Code
Date of Birth	Weight（kg）		Height（cm）
Working Experience			

Ⅳ. Complete the following dialogue and read it aloud.

A: May I have your name?

B: Yes. _____.

A: How old are you?

B: _____.

A: What's your major?

B: _____.
A: What's your hobby?
B: _____.
A: Why do you want to work for our hotel?
B: _____.
A: Do you think you're qualified for this job?
B: I think I am. _____.
A: Do you have any part-time work experience?
B: _____.

V. Translate the following sentences into English.
1. 你是做什么的？我是一名厨师。
2. 我是一名早餐厨师，负责做早餐。
3. 你认为你的工作怎么样？
4. 我能加班工作并且能和他人愉快地相处。
5. 我性格外向。

Part D: Reading

A Letter of Application

15 Lijia Street
Dalian, 116001
15th September, 2013

Furama Hotel
Renmin Road
Dalian, 116012
Dear sir or madam,

    I learned about your job vacancy from the advertisement on today's newspaper.

    It's said that your hotel wants to employ cook. I take it as a good opportunity, and here are the qualifications I can offer.

    I'm 20 years old. I'm an easy-going girl. I'm always helpful. I enjoy offering my help if others need. I can get along well with people.

    I'm a vocational school graduate. My major is Chinese Pastry. I once worked in a restaurant as a trainee last summer. There I practiced what I've learned at school, such as making dumplings, steamed buns, noodles and so on. I cooperated with my colleagues well. I don't mind hard work and long working hours. Please consider me for the job.

 烹饪厨房英语

I enclose a card with my address and telephone number, in the hope you can use it to inform me of the interview.

Faithfully yours,
Wang Fang

Words & Expressions

vacancy ['veikənsi] *n.*　空缺
employ [im'plɔi] *vt.*　雇佣，使用
easy-going [ˌi:zi:'gəuiŋ] *adj.*　随和的
major ['meidʒə（r）] *n.*　主修科目

advertisement [əd'vɜ:tismənt] *n.*　广告
qualification [ˌkwɔlifi'keiʃn] *n.*　资格
opportunity [ˌɔpə'tju:nəti] *n.*　机会
interview ['intəvju:] *n.*　面试

Fruit（水果）

pineapple 菠萝
strawberry 草莓
blueberry 蓝莓
mango 杧果
cherry 樱桃
litchi 荔枝
orange 橙子
banana 香蕉
apple 苹果
watermelon 西瓜
grape 葡萄
pear 梨
apricot 杏
peach 桃
date 枣
mangosteen 山竹
pomegranate 石榴
pomelo 柚子
hawthorn 山楂
olive 橄榄
kiwi 猕猴桃
papaya 木瓜
sugarcane 甘蔗
wax apple 莲雾
waxberry 杨梅
carambola 阳桃
betelnut 槟榔
durian 榴梿
coconut 椰子
longan 龙眼，桂圆
fig 无花果
muskmelon 香瓜
plum 李子
avocado 牛油果
persimmon 柿子
pitaya 火龙果
fig 无花果
loquat 枇杷
guava 番石榴

Vegetable（蔬菜）

asparagus 芦笋
broccoli 西兰花
cabbage 卷心菜
carrot 胡萝卜
cucumber 黄瓜
celery 芹菜
Chinese cabbage 大白菜
eggplant 茄子
leek 韭菜

烹饪厨房英语

lettuce　生菜
mushroom　蘑菇
onion　洋葱
spinach　菠菜
lotus root　莲藕
green pepper　青椒
potato　土豆
pumpkin　南瓜
scallion　葱
tomato　西红柿
pea　豌豆
taro　芋头
white gourd　冬瓜
rape　油菜
caraway　香菜
bamboo shoot　竹笋
fennel　茴香
needle mushroom　金针菇
agar　紫菜
radish　萝卜
lentil　小扁豆
sweet potato　红薯，甘薯
parsley　欧芹
mung bean　绿豆
chive　香葱
ginger　生姜
yam　山药
lily　百合
marrow　西葫芦
garlic　蒜
winter bamboo shoots　冬笋
bitter gourd　苦瓜
cauliflower　花菜
kale　甘蓝菜
bean curd　豆腐
kidney bean　四季豆

Meat & Poultry（肉和家禽）

pork　猪肉
beef　牛肉
mutton　羊肉
lamb　羔羊肉
lean meat　瘦肉
speck　肥肉
streaky pork　五花肉
spareribs　小排骨
steak　牛排
smoked bacon　熏肉
goose liver　鹅肝
sausage　香肠
chicken　鸡肉
turkey　火鸡
duck　鸭
goose　鹅
roast duck　烤鸭
chicken wing　鸡翅
chicken claw　凤爪
chicken breasts　鸡胸脯肉
black pudding　血肠
OX-tail　牛尾
lard　猪油
pig kidney　猪腰
gizzard　鸡胗
pig's feet　猪蹄
pork fillet　小里脊肉
backstraps　板筋
tenderloin　牛柳，里脊肉
pork ribs　肋骨
pig's liver　猪肝
OX-tongues　牛舌
veal　小牛肉
lard　猪油
ham　火腿

Words

ground pork　绞肉

Seafood（海鲜）

cod　鳕鱼
sea perch　海鲈鱼
salmon　三文鱼
trout　鳟鱼
tuna　金枪鱼
mandarin fish　鳜鱼
yellow croaker　黄花鱼
hairtail　带鱼
crucian carp　鲫鱼
plaice　鲽鱼
grass carp　草鱼
spanish mackerel　鲅鱼
sole　舌鳎
flounder　比目鱼
halibut　大比目鱼
eel　鳗鱼
sardine　沙丁鱼
wolf fish　海鲶鱼
abalone　鲍鱼
conch　海螺
clam　蛤蚌
oyster　牡蛎
scallop　扇贝
squid　鱿鱼
octopus　章鱼
cuttlefish　墨鱼
lobster　龙虾
prawn　对虾
shrimp　小虾，河虾
carp　鲤鱼
crab　螃蟹
catfish　鲶鱼
herring　鲱鱼
winkle　田螺
sea urchin　海胆
sea cucumber　海参

Dairy Products（乳制品）

whole milk　全脂牛奶
low-fat milk　低脂牛奶
skim milk　脱脂牛奶
sherbet　冰糕
condensed milk　炼乳
regular cream　普通奶油
heavy cream　浓奶油
sour cream　酸奶油
yogurt　酸奶
ice cream　冰激凌
butter　黄油
cheese　奶酪

Drinks（饮品）

beverage　酒水
red wine　红葡萄酒
white wine　白葡萄酒
whisky　威士忌
champagne　香槟
tea　茶
soda　苏打水
ginger ale　干姜水
7-UP　七喜
Coca Cola　可口可乐
Pepsi Cola　百事可乐
beer　啤酒
samshu　烧酒
Sake　日本清酒
rice wine　米酒
milk　牛奶
coffee　咖啡

milk shake   奶昔
fruit juice   果汁
mineral water   矿泉水
apple cider vinegar   苹果醋
black tea   红茶
green tea   绿茶
rum   朗姆酒
gin   杜松子酒
vodka   伏特加
tequila   龙舌兰
distill water   蒸馏水
aperitifs   开胃酒
brandy   白兰地
sherry   雪利酒
cocktail and mixed drinks   鸡尾酒
Fanta   芬达
red bull vitamin drinks   红牛

Seasoning（调味品）

oil   油
soy sauce   酱油
brown sauce   布朗沙司
cream sauce   奶油沙司
red wine sauce   红酒沙司
rosemary sauce   迷迭香沙司
orange sauce   鲜橙沙司
pepper sauce   胡椒沙司
Hollandaise sauce   荷兰沙司
French dressing   法国汁
curry sauce   咖喱沙司
Paris butter   巴黎黄油
snail butter   蜗牛黄油
mayonnaise sauce   马乃司沙司
tartar sauce   鞑靼沙司
caviar sauce   鱼子酱沙司
tomato sauce   番茄沙司

thousand islands dressing   千岛汁
ketchup   番茄酱
salt   盐
MSG   味精
chicken essence   鸡精
sesame oil   芝麻油
chili oil   辣椒油
white sugar   白糖
brown sugar   红糖
rock sugar   冰糖
cooking wine   料酒
starch   淀粉
soybean paste   大豆酱
scallion   葱
ginger   姜
garlic   蒜
pepper   胡椒粉
star anise   八角
caraway   香菜
cumin   孜然
curry   咖喱
cinnamon   肉桂
caviar   鱼子酱
sweet soybean paste   甜面酱
mustard   芥末
honey   蜂蜜
dried orange peel   陈皮
cream   奶油
cocoa powder   可可粉
jam   果酱
soy sauce   酱油
yeast   酵母
vinegar   醋
vanilla   香草
thyme   百里香
laurel   月桂

Words

sesame oil　芝麻油
sauce　调味汁
rosemary　迷迭香
peppermint　薄荷
clove　丁香
sage　鼠尾草
basil　罗勒
dill　莳萝
saffron　藏红花
marjoram　牛膝草
oregano　小茴香
tarragon　龙蒿
citronella　香茅
paprika　红椒粉
bay leaf　香叶

Chinese Pastry（中式面点）

steamed stuffed bun　包子
boiled dumpling　水饺
steamed dumpling　蒸饺
wonton　馄饨
noodle　面条
handmade noodle　手擀面
hand-pulled noodle soup　拉面
steamed bread　馒头
steamed roll　花卷
triangle bun stuffed with sugar　糖三角
fried dumplings　煎饺
fried leek dumplings　韭菜盒子
deep-fried pancake　油饼
baked scallion pancake　葱油饼
deep-fried glutinous rice balls with sesame　炸麻球
twist of dough　麻花
deep-fried dough stick　油条
Beijing-style cakes　京式糕点

Tangyuan（Glutinous Rice Balls）　汤圆
small steamed bun　小笼包子
spring pancake　春饼
sweetheart cake　老婆饼
moon cake　月饼
bean paste cake　绿豆糕
rice cake　年糕
leaf-fat bun　水晶包
smashed bean bun　豆沙包
baozi stuffed with creamy custard　奶黄包
glutinous rice rolls stuffed with red bean paste　驴打滚
peach-shaped mantou　寿桃
baked wheat cake　火烧
glutinous rice rolls　糯米卷
dough drop and assorted vegetable soup　疙瘩汤
Cantonese Dim Sum　广东点心
lotus seed puff pastry　莲蓉酥
baozi stuffed with BBQ pork　叉烧包
pan-fried baozi stuffed with pork　生煎包
crispy durian pastry　榴梿酥
honey BBQ pork puff　蜜汁叉烧酥
pumpkin puff　南瓜酥
red bean cake　红豆糕
spring rolls　春卷
baked corn pancake　小贴饼子
minced pork pancake　肉松饼
potato pancake　土豆饼
noodles in chili sauce, Sichuan style　担担面

Western Pastry（西式面点）

pastry　油酥面包
toast　吐司，面包片
yeast bread　酵母面包
black bread　黑面包

hard roll 硬质面包
soft roll 软质面包
muffin 松饼
cake 蛋糕
cupcake 小蛋糕
cream cake 奶油蛋糕
cheese cake 奶酪蛋糕
pound cake 磅蛋糕
angel food cake 天使蛋糕
pie 派
banana cream pie 香蕉奶油派
cherry pie 樱桃派
chocolate cream pie 巧克力奶油派
apple pie 苹果派
walnut pie 核桃派
pumpkin pie 南瓜派
tart 蛋挞
cookie 曲奇
biscuit 饼干
bavarian cream 巴伐利亚奶油
ice cream 冰激凌
chocolate mousse 巧克力慕斯
green tea mousse 绿茶慕斯
cream mousse 奶油慕斯
ginger cake 姜饼
pancake 煎饼
mousse 慕斯
doughnut 甜甜圈
pudding 布丁
waffle 华夫饼
sponge cake 海绵蛋糕

Chinese Cuisine（中餐）

Meat（肉类）

fried pork mixed with sweet and sour sauce 糖醋里脊
roasted suckling pig 烤乳猪
braised pork leg 红烧猪蹄
sauté pork cubelets with hot pepper 宫保肉丁
braised pork tendons 红烧蹄筋
meat balls braised with brown sauce 红烧狮子头
Nanjing pressed duck 南京板鸭
stewed chicken with mushroom 小鸡炖蘑菇
braised pork with vermicelli 猪肉炖粉条
boiled salted duck 盐水鸭
bonbon chicken 棒棒鸡
pork lungs in chili sauce 夫妻肺片
sweet and sour pork 咕噜肉
sweet and sour spareribs 糖醋排骨
pork with preserved vegetable 梅菜扣肉
quick-fried pork and scallions 葱爆肉
double cooked pork slices 回锅肉
chicken with chili and pepper 辣子鸡
fish-flavored shredded pork with chili（in Sichuan style） 鱼香肉丝
curried beef 咖喱牛肉
curried chicken 咖喱鸡
braised chicken 焖鸡

Seafood（海鲜）

braised carp 红烧鲤鱼
braised sea cucumbers in brown sauce 红烧海参
braised prawns 油焖大虾
steamed minced pork with salt fish 咸鱼蒸肉饼
fried shrimps with Longjing Tea 龙井虾仁
fish balls in clear soup 清汤鱼丸
stewed soft-shelled turtle with ham 火腿炖甲鱼
deep fried squid 酥炸鲜鱿

Words

boiled prawns　白灼虾
steamed sea grouper　清蒸石斑
scallop with garlic sauce　鱼香干贝
braised cat fish with eggplant　鲶鱼烧茄子
steamed fish head with chilli pepper　剁椒鱼头
boiled fish with Sichuan pickles　酸菜鱼

Vegetables（蔬菜）

sautéed sweet corn with salty egg yolk　黄金玉米
sautéed black fungus with celery　芹香木耳
sautéed green chili pepper　虎皮尖椒
mapo tofu　麻婆豆腐
hot candied sweet potato　拔丝地瓜
sautéed corn with pine seeds　松仁玉米
grilled assorted vegetables　铁扒什锦
eggplants with garlic sauce　鱼香茄子
hot spicy chinese cabbage　辣白菜

Soup（汤）

egg drop soup　蛋花汤
three-delicacy soup　三鲜汤
noodles soup　汤面
hot & vegetable soup　酸辣汤
oyster soup　牡蛎汤
seaweed soup　紫菜汤
carp soup　鲫鱼汤
turtle soup with ham　火腿甲鱼汤
mashed chicken and asparagus in soup　芦笋鸡肉汤

Western Cuisine（西餐）

buffet　冷餐

cold dish 冷菜

salad　沙拉
ham salad　火腿沙拉
vegetable salad　蔬菜沙拉
meat　肉
cold mixed meat　冷什锦肉
cold meat and sausage　冷肉拼香肠
mashed liver, live paste　肝泥
cold roast beef　冷烤牛肉
cold stewed sausage　冷烩茶肠
fish　鱼
stewed fish slices with brown sauce　红烩鱼片
stewed fish slices with tomato sauce　茄汁烩鱼片
minced herring with eggs　鸡蛋鲱鱼泥子
smoked herring　熏鲱鱼
sardine　沙丁鱼
minced prawns　大虾泥
minced crab meat　蟹肉泥
poultry　家禽
chicken jelly, chicken in aspic　鸡肉冻
minced chicken meat, chicken paste　鸡肉泥
minced chicken liver, chicken liver paste　鸡肝泥
minced duck liver, duck liver paste　鸭肝泥
cold roast chicken with vegetables　冷烤油鸡蔬菜
vegetable dish　素菜
assorted vegetables　什锦蔬菜
stewed egg-plant brown sauce　红烩茄子
sour cucumbers, pickled cucumbers　酸黄瓜
pickled cabbage, sour and sweet cabbage　泡菜

appetizers 热小菜

stewed sausage with cream　奶油烩香肠
fried herring with egg sauce　鸡蛋汁煎鲱鱼
fried eggs　煎鸡蛋
fried eggs with ham, ham and eggs　火腿煎蛋

烹饪厨房英语

fried eggs with bacon, bacon and eggs　咸肉煎蛋
fried eggs with sausage, sausage and eggs　香肠煎蛋
omelette/omelet　煎蛋卷
sausage omelette/omelet　香肠炒蛋
ham omelette/omelet　火腿蛋饼

soup 汤

light soup, clear soup, consomme　清汤
thick soup, potage　浓汤
broth　肉汤
creamed ham soup, ham soup with cream　奶油火腿汤
creamed mashed chicken soup, mashed chicken soup with cream　奶油鸡茸汤
creamed crab meat soup, crab meat soup with cream　奶油蟹肉汤
creamed mushroom soup with crab meat　奶油口蘑蟹肉汤
mushroom soup with cauliflower　奶油口蘑花菜汤
creamed tomato soup, tomato soup with cream　奶油西红柿汤
creamed peas soup, peas soup with cream　奶油豌豆汤
creamed mashed peas soup, mashed peas soup with cream　奶油豌豆泥汤
mixed meat soup　肉杂拌汤
OX-tail soup　牛尾汤
beef balls soup　牛肉丸子汤
beef soup with vegetables　牛肉蔬菜汤
beef tea　牛肉茶
cold beef tea　冷牛肉茶
chicken soup　鸡汤
chicken soup with mushrooms　口蘑鸡汤
chicken soup with tomato　番茄鸡汤
chicken leg soup with vegetables　鸡腿蔬菜汤
curry chicken cubes soup　咖喱鸡丁汤
fish soup　鱼汤
fish soup with tomato　红鱼汤
soup with vegetables　蔬菜汤
pickled vegetable soup　酸菜汤
onion soup　洋葱汤
tomato soup　西红柿汤
white bean soup　白豆汤
pea soup　豌豆汤
consomme with meat pie　清汤肉饼
soup with macaroni　通心粉汤
consomme with macaroni　通心粉清汤
tomato soup with macaroni　番茄通心粉汤
fried mandarin fish　炸鳜鱼
fried fish with ham sauce　火腿汁煎鱼
grilled mandarin fish　铁扒鳜鱼
baked fish with cream sauce　奶油汁烤鱼
fish steak　鱼排
mandarin fish au gratin　奶油口蘑烤鳜鱼
steamed fish with red wine　红酒蒸鱼
steamed fish with lemon sauce　柠檬汁蒸鱼
fried prawns　炸大虾
soft-fried prawns　软煎大虾

vegetable dish 素菜

cauliflower with butter　黄油花菜
mixed vegetables with butter　黄油杂拌蔬菜
fried mushrooms with butter　黄油炒口蘑
fried spinach with butter　黄油炒菠菜
fried peas with butter　黄油炒豌豆
fried green peas with butter　黄油炒青豆
fried eggplant slices　炸茄子片
fried tomato　炸番茄
fried potato cake　清煎土豆饼

Words

braised cabbage rolls　家常焖洋白菜卷
stewed eggplants　烩茄子
omelette/omelet with green beans　扁豆炒蛋
curry vegetables　咖喱素菜
roast chicken　烤鸡
roast chicken with vegetables　素菜烤鸡
fried chicken　煎鸡
eep-fried chicken　炸鸡
boiled chicken with macaroni　通心粉煮鸡
boiled chicken with cream sauce　奶煮鸡
grilled spring chicken　铁扒笋鸡
braised chicken　焖鸡
braised chicken with butter　黄油焖鸡

[1] 周海霞，章敏均.烹饪英语[M].北京：科学出版社，2011.
[2] 吴蓓蓓，等.餐饮服务行业实用英语对话及词汇手册[M].北京：中国水利水电出版社，2009.
[3] 姜玲.厨师职业英语[M].北京：高等教育出版社，2009.
[4] 姜玲.厨师岗位英语[M].北京：旅游教育出版社，2008.
[5] 金惠康，罗向阳.酒楼服务英语[M].广州：广东旅游出版社，2006.
[6] 杜纲.烹饪英语[M].重庆：重庆大学出版社，2013.
[7] 张艳红.中餐烹饪英语[M].重庆：重庆大学出版社，2015.
[8] 宋洁.烹饪英语[M].2版.北京：中国轻工业出版社，2020.
[9] 蔡琳琳.西餐烹饪英语[M].4版.北京：旅游教育出版社，2019.